ENCYCLOPEDIA OF ENVIRONMENTAL ISSUES

# ATMOSPHERE AND
# AIR POLLUTION

# ENCYCLOPEDIA OF ENVIRONMENTAL ISSUES
# ATMOSPHERE AND AIR POLLUTION

*Editor*

**Craig W. Allin**

*Cornell College*

SALEM PRESS

A Division of EBSCO Publishing, Ipswich, Massachusetts

Cover photo:
*Heavy smog over Tomorrow Square in downtown Shanghai.* (© Keren Su/Corbis)

ISBN: 978-1-42983-670-8

# Table of Contents

# Contributors

Richard Adler
*University of Michigan-Dearborn*

Anita Baker-Blocker
*Ann Arbor, Michigan*

Grace A. Banks
*Chestnut Hill College*

Raymond D. Benge, Jr.
*Tarrant County College-Northeast Campus*

Massimo D. Bezoari
*Huntingdon College*

Cynthia A. Bily
*Macomb Community College*

Margaret F. Boorstein
*C. W. Post College of Long Island University*

Kenneth H. Brown
*Northwestern Oklahoma State University*

Jeffrey C. Brunskill
*Bloomsburg University of Pennsylvania*

Michael A. Buratovich
*Spring Arbor University*

Byron Cannon
*University of Utah*

Jeff Cervantez
*University of Tennessee*

Dennis W. Cheek
*Ewing Marion Kauffman Foundation*

Thomas Clarkin
*University of Texas at San Antonio*

Daniel J. Connell
*Boston College*

Mark Coyne
*University of Kentucky*

Ralph D. Cross
*University of Southern Mississippi*

John M. Dunn
*Ocala, Florida*

Robert D. Engelken
*Arkansas State University*

George J. Flynn
*State University of New York at Plattsburgh*

Soraya Ghayourmanesh
*Bayside, New York*

Craig S. Gilman
*Coastal Carolina University*

D. R. Gossett
*Louisiana State University Shreveport*

Daniel G. Graetzer
*University of Washington Medical Center*

William Crawford Green
*Morehead State University*

Phillip A. Greenberg
*San Francisco, California*

Wendy Halpin Hallows
*Chestnut Hill College*

Laurent Hodges
*Iowa State University*

Ronald K. Huch
*University of Papua New Guinea*

Diane White Husic
*East Stroudsburg University*

Bernard Jacobson
*SciMed Writers*

Jeffrey A. Joens
*Florida International University*

Bruce E. Johansen
*University of Nebraska at Omaha*

Karen N. Kähler
*Pasadena, California*

Grove Koger
*Boise State University*

Timothy Lane
*Louisville, Kentucky*

Thomas T. Lewis
*Mount Senario College*

David C. Lukowitz
*Hamline University*

Fai Ma
*University of California, Berkeley*

Francis P. Mac Kay
*Providence College*

Louise Magoon
*Fort Wayne, Indiana*

M. Marian Mustoe
*Eastern Oregon University*

Alice Myers
*Bard College at Simon's Rock*

Anthony J. Nicastro
*West Chester University*

Beth Ann Parker
*Huntingdon College*

George R. Plitnik
*Frostburg State University*

Noreen D. Poor
*University of South Florida*

Gene D. Robinson
*James Madison University*

Charles W. Rogers
*Southwestern Oklahoma State University*

Joseph R. Rudolph, Jr.
*Towson University*

Alexander Scott
*Pasadena, California*

Martha A. Sherwood
*Kent Anderson Law Office*

Paul P. Sipiera
*Harper College*

Alexander R. Stine
*University of California, Berkeley*

Rena Christina Tabata
*University of British Columbia*

John M. Theilmann
*Converse College*

Oluseyi A. Vanderpuye
*Albany State University*

Shawncey Webb
*Taylor University*

Kay R. S. Williams
*Shippensburg University*

Marcie L. Wingfield
*Huntingdon College*

Brian G. Wolff
*Minnesota State Colleges and Universities*

Robin L. Wulffson
*Faculty, American College of Obstetrics and Gynecology*

# Acid deposition and acid rain

CATEGORY: Atmosphere and air pollution

DEFINITION: Deposition of acidic gases, particles, and precipitation (rain, fog, dew, snow, or sleet) on the surface of the earth

SIGNIFICANCE: Electric utilities, industries, and automobiles emit sulfur dioxide and nitrogen oxides that are readily oxidized into sulfuric and nitric acids in the atmosphere. Long-range transport and dispersion of these air pollutants produce regional acid deposition, which alters aquatic and forest ecosystems and accelerates corrosion of buildings, monuments, and statuary.

In 1872 Robert Angus Smith used the term "acid rain" in his book *Air and Rain: The Beginnings of a Chemical Climatology* to describe precipitation affected by coal-burning industries. Acidity is created when sulfur dioxide ($SO_2$) and nitrogen oxides ($NO_x$) react with water and oxidants in the atmosphere to form water-soluble sulfuric and nitric acids. The normal acidity of rain is pH 5.6, which is caused by the formation of carbonic acid from water-dissolved carbon dioxide. The acidity of precipitation collected at monitoring stations around the world varies from pH 3.8 to 6.3 (pH 3.8 is three hundred times as acidic as pH 6.3). Ammonia, as well as soil constituents such as calcium and magnesium that are often present in suspended dust, neutralizes atmospheric acids, which helps explain the geographical variation of precipitation acidity.

## INCREASING ACIDITY

Between the mid-nineteenth century and World War II, the Industrial Revolution led to a tremendous increase in coal burning and metal ore processing in both Europe and North America. The combustion of coal, which contains an average of 1.5 percent sulfur by weight, and the smelting of metal sulfides released opaque plumes of smoke and $SO_2$ from short chimneys into the atmosphere.

Copper, nickel, and zinc smelters inundated nearby landscapes with $SO_2$ and heavy metals. One of the world's largest nickel smelters, located in Sudbury, Ontario, Canada, began operation in 1890 and by 1960 was pouring 2.6 million tons of $SO_2$ per year into the atmosphere. By 1970 the environmental damage extended to 72,000 hectares (278 square miles) of injured vegetation, lakes, and soils sur-

rounding the site; within this area 17,000 hectares (66 square miles) were barren. The land was devastated not only by acid deposition but also by the accumulation of toxic metals in the soil, the clear-cutting of forested areas for fuel, and soil erosion caused by wind, water, and frost heave. (The situation improved dramatically in the ensuing decades, but only after construction in 1972 of a tall "superstack" that dispersed emissions farther from the smelting facility, followed by an extensive tree-planting program and other major environmental reclamation efforts, as well as installation of industrial scrubber systems at the facility during the 1990's.)

In urban areas, high concentrations of sulfur corroded metal and accelerated the erosion of stone buildings and monuments. Structures such as the Acropolis in Greece suffered serious damage from elevated acidity. During the winter, added emissions from home heating and stagnant weather conditions caused severe air pollution episodes characterized by sulfuric acid fogs and thick, black soot. In 1952 a four-day air-pollution episode in London, England, killed an estimated four thousand people.

After World War II, large coal-burning utilities in Western Europe and the United States built their plants with particulate control devices and stacks higher than 100 meters (328 feet) to improve the local air quality. (By contrast, huge industrial facilities throughout Eastern Europe and the Soviet Union operated without air-pollution controls for most of the twentieth century.) The tall stacks increased the dispersion and transport of air pollutants from tens to hundreds of kilometers. While this measure eased localized impacts, it simultaneously made the problem more widespread. Worldwide emissions of $SO_2$ increased; in the United States emissions climbed from 18 million tons in 1940 to a peak of 28 million tons in 1970. Acid deposition evolved into an interstate and even an international problem.

In major cities, exhaust from automobiles combined with power plant and industrial emissions to create a choking, acrid smog of ozone mixed with nitric and organic acids formed by photochemical processes. The rapid deterioration of air quality in cities, with attendant health and environmental consequences, spurred the passage of environmental laws such as the U.S. Clean Air Act (CAA) of 1963, which was amended and expanded in 1970, 1977, and 1990. Each amendment to the CAA brought new requirements for air-pollution controls.

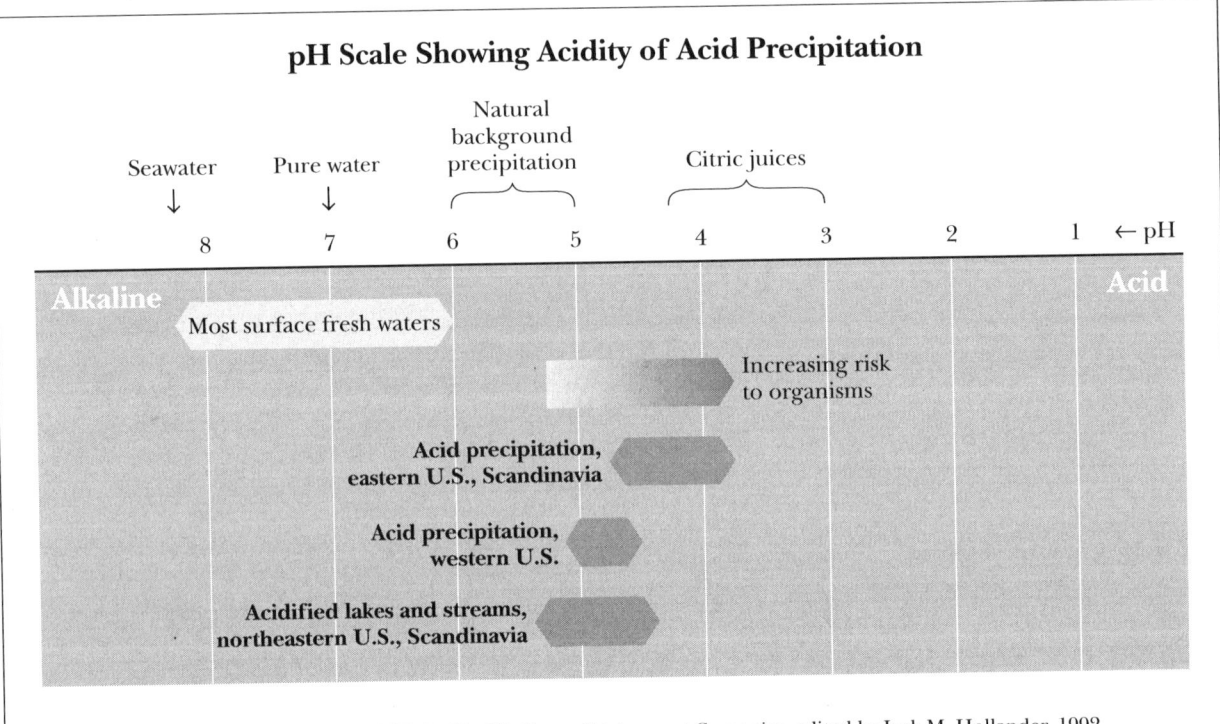

**pH Scale Showing Acidity of Acid Precipitation**

*Source:* Adapted from John Harte, "Acid Rain," in *The Energy-Environment Connection*, edited by Jack M. Hollander, 1992.
*Note:* The acid precipitation pH ranges given correspond to volume-weighted annual averages of weekly samples.

EFFECTS ON AQUATIC ECOSYSTEMS

Landscapes rich in limestone or acid-buffering soils are less sensitive to acid deposition. Regions that are both sensitive and exposed to acid deposition include the eastern United States, southeastern Canada, the northern tip of South America, southern Sweden and Norway, central and Eastern Europe, the United Kingdom, southeastern China, northeastern India, Thailand, and the Republic of Korea (South Korea). Within these regions, acid rain disrupts aquatic ecosystems and contributes to forest decline.

A strong correlation has been found between fish extinction and lake and stream acidity. Researchers have also found that the diversity of not only fish but also phytoplankton, zooplankton, invertebrates, and amphibian species diminishes by more than 50 percent as surface water pH drops from 6.0 to 5.0. Below pH 5.6, aluminum released from bottom sediments or leached from the surrounding soils interferes with gas and ion exchange in fish gills and can be toxic to aquatic life. At pH 5.0, most fish eggs cannot hatch; below pH 4.0, no fish survive.

In southern Norway, the virtual extinction of salmon in all the larger salmon rivers is attributed to acid rain. In 2008 it was found that twelve salmon stocks were endangered and another eighteen had been wiped out altogether. Records and long-term monitoring show that the decline of fish populations began during the early twentieth century, with dramatic losses during the 1950's. Between 1950 and 1990, fish mortality in the region became more widespread. In the three decades following 1980, pollution-control measures in Norway and the rest of Europe cut acid precipitation over Norway by roughly half. Aquatic animal and plant populations and ecosystems continue to recover from the damage.

In the United States the New York State Department of Environmental Conservation reported in 2008 that 26 percent of the state's lakes in the Adirondack Mountains were unable to neutralize incoming acids to concentrations that fish could tolerate. During certain times of year, up to 70 percent of the lakes had the potential to become intolerably acidic. Of forty-eight Adirondack lakes surveyed, sixteen had aluminum concentrations above levels that juvenile fish could withstand. Ecosystems in this region have been affected by plumes of air pollutants carried by prevailing winds from the Ohio River Valley. Like their Norwegian counterparts, the Adirondack lakes have low acid-neutralizing capacity. Fish declines that began during the early twentieth century and continued through the 1980's corresponded to reductions in pH. Fish kills often followed spring snowmelt, which filled the waterways with acid accumulated in winter precipitation.

While the CAA and its amendments have done much to reduce the acidity of the precipitation in the Adirondacks, and thus of the lakes, recovery takes time. Lake chemistry that is acidic year-round may take twenty-five to one hundred years to return to levels that aquatic life can tolerate, while seasonally acidic lake waters may take a few years or several decades. Once a lake regains its ability to neutralize acid, wildlife in the lake's ecosystem require additional time to become reestablished.

## EFFECTS ON FORESTS AND CITIES

In areas exposed to acid rain, dead and dying trees stand as symbols of environmental change. In Germany the term *Waldsterben*, or forest death, was coined to describe the rapid declines of Norway spruce, Scotch pine, and silver fir trees during the early 1980's, followed by beech and oak trees during the late 1980's, especially at high elevations in the Black and Bavarian forests. At higher altitudes, clouds frequently shroud mountain peaks, bathing the forest canopy in a mist of heavy metals and sulfuric and nitric acids. Under drought conditions, invisible plumes of ozone from sources hundreds of kilometers distant intercepted the mountain slopes. Several forests within the United States have been likewise affected by both ozone and acidic deposition, including the pine forests in California's southern Sierra Nevada and high-elevation spruce-fir forests in the Northeast.

Intensive field and laboratory investigations of forest decline in North America and Europe have yielded contradictory results regarding the link between dead trees and acid deposition. Some laboratory experiments have found that acid rain has no effect, or even a fertilizing effect, on trees. Symptoms of tree stress include changes in foliage color, size, and shape; destruction of fine roots and associated fungi; and stunted growth. Many researchers attribute these symptoms and forest decline to complex interactions among a variety of stressors, including acid precipita-

tion, ozone, excessive nitrogen deposition, land management practices, climate change, drought, and pestilence. Acid rain generally does not kill trees directly; rather, it weakens them by damaging their leaves, impairing their leaves' functionality, stripping nutrients from the soil, and releasing toxic substances from the soil.

Ambient air concentrations of $SO_2$ and $NO_x$ are typically higher in major cities owing to the high density of emission sources in these locations. The resulting haze may be carried by winds hundreds of kilometers from where it was generated, obscuring visibility even in remote wilderness areas. The acids formed accelerate the weathering of exposed stone, brick, concrete, glass, metal, and paint. For example, the calcite in limestone and marble reacts with water and sulfuric acid to form gypsum (calcium sulfate). The gypsum washes off stone with rain or, if eaves protect the stone, accumulates as a soot-darkened crust. Acid-induced weathering has obscured the details of elaborate carvings on medieval cathedrals, ancient Greek columns, and Mayan ruins at alarming rates.

## PREVENTION EFFORTS

Discovery of the connection between sulfur emissions in continental Europe and lake acidification in Scandinavian countries ultimately led to the 1979 Convention on Long-Range Transboundary Air Pollution. This legally binding international agreement addresses air-pollution issues on a broad regional basis. Subsequent protocols to the convention deal specifically with sulfur emissions (1985 and 1994 protocols), $NO_x$ emissions (1988 protocol), and acidification abatement (1999 protocol).

In the United States the Acidic Deposition Control Program, Title IV of the CAA amendments of 1990, directs the Environmental Protection Agency (EPA) to reduce the adverse effects of acid rain, specifically through a reduction in emissions of $SO_2$ and $NO_x$. The program sets an annual cap on $SO_2$ emissions from power plants and establishes allowable $NO_x$ emission rates based on boiler type. Between 1990 and 2006, the program achieved a reduction of $SO_2$ emissions by more than 6.3 million tons from 1990 levels. The program, in combination with other efforts to reduce emissions, also cut $NO_x$ emission by roughly 3 million tons, making 2006 emissions less than half what would have been expected without the program.

The National Acid Precipitation Assessment Program coordinates interagency acid deposition monitoring and research and assesses the cost, benefits, and effectiveness of acid deposition control strategies. Acid deposition reduction schemes in the United States target large electric utilities, which as of 2006 were responsible for about 67 percent of the country's $SO_2$ and 19 percent of the $NO_x$ emissions from anthropogenic sources. Utilities participate in a novel market-based emission allowance trading and banking system that permits great flexibility in controlling $SO_2$ emissions. For example, utilities may choose to remove sulfur from coal by cleaning it, to burn a cleaner fuel such as natural gas, or to install a gas desulfurization system to reduce emissions. They may also buy or sell emissions allowances. U.S. research efforts cost-shared between government and industry, such as the Clean Coal Power Initiative and similar programs before it, have developed technologies—for example, the catalytic conversion of $NO_x$ to inert nitrogen—that can radically decrease emissions of acid gases from coal-fired power plants.

Between 1990 and 2006 annual atmospheric concentrations of $SO_2$ in the United States decreased by 53 percent, while nitrogen dioxide concentrations fell by 30 percent. Emissions of $SO_2$ dropped during this period by 38 percent; $NO_x$ emissions declined 29 percent, with most of the decrease occurring after 1998. These reductions resulted in significant decreases in acid rain. Between 1985 and 2002, nitrate deposition declined in the New England states; however, in the western states, increasing oil and gas production and other factors caused nitrate deposition to rise. In the eastern states, sulfate concentrations in the air generally declined, affected regions decreased in size, and the magnitude of the highest concentrations dropped. During the periods from 1989 to 1991 and 2004 to 2006, sulfate deposition in the Northeast and the Midwest decreased by more than 30 percent. As a result, surface water quality improved.

*Noreen D. Poor*
*Updated by Karen N. Kähler*

FURTHER READING

Brimblecombe, Peter, et al., eds. *Acid Rain: Deposition to Recovery.* New York: Springer, 2007.

Finlayson-Pitts, Barbara J., and James N. Pitts. *Chemistry of the Upper and Lower Atmosphere: Theory, Experiments, and Applications.* San Diego, Calif.: Academic Press, 2000.

Hill, Marquita K. "Acidic Deposition." In *Understanding Environmental Pollution.* 3d ed. New York: Cambridge University Press, 2010.

Lane, Carter N., ed. *Acid Rain: Overview and Abstracts.* New York: Nova Science, 2003.

Little, Charles E. *The Dying of the Trees: The Pandemic in America's Forests.* New York: Viking Press, 1995.

McGee, Elaine. *Acid Rain and Our Nation's Capital: A Guide to Effects on Buildings and Monuments.* Washington, D.C.: U.S. Geological Survey, 1995.

Smith, Robert Angus. *Air and Rain: The Beginnings of a Chemical Climatology.* 1872. Reprint. Whitefish, Mont.: Kessinger, 2007.

Visgilio, Gerald R., and Diana M. Whitelaw, eds. *Acid in the Environment: Lessons Learned and Future Prospects.* New York: Springer, 2007.

# Aerosols

CATEGORY: Atmosphere and air pollution

DEFINITION: Aggregations of small particles, both liquid and solid, and the atmospheric gases in which they are suspended

SIGNIFICANCE: Atmospheric aerosols are responsible for the diminishment of environmental air quality throughout much of the world.

An aerosol is a multiphasic system consisting of tiny liquid and solid particles and the gas in which they are suspended. In unpolluted areas such as New Zealand, aerosols contain impurities from natural sources; acidity comes from carbonic acid ($H_2CO_3$). In central Europe and other industrialized areas throughout the world, fossil-fuel combustion contributes large amounts of oxides of sulfur ($SO_x$) and oxides of nitrogen ($NO_x$) to the atmosphere, leading to the formation of sulfuric acid and nitric acid aerosols. In these polluted areas, acidity levels are much higher in fog than in rain, and dry deposition of sulfuric and nitric acid particulates from impactions of acid fog may be more damaging to buildings and the environment than acid rain. Forest canopies tend to scavenge acid aerosols; on conifer-covered mountains, cloud droplets are the major source of acid deposition. Dry deposition on canopies rapidly affects root systems and increases soil acidification over the long term. Soil acidification has been linked to a number of adverse effects on vegetation, especially tree dieback and forest decline in Europe.

## Radiative Forcing by Tropospheric Aerosol

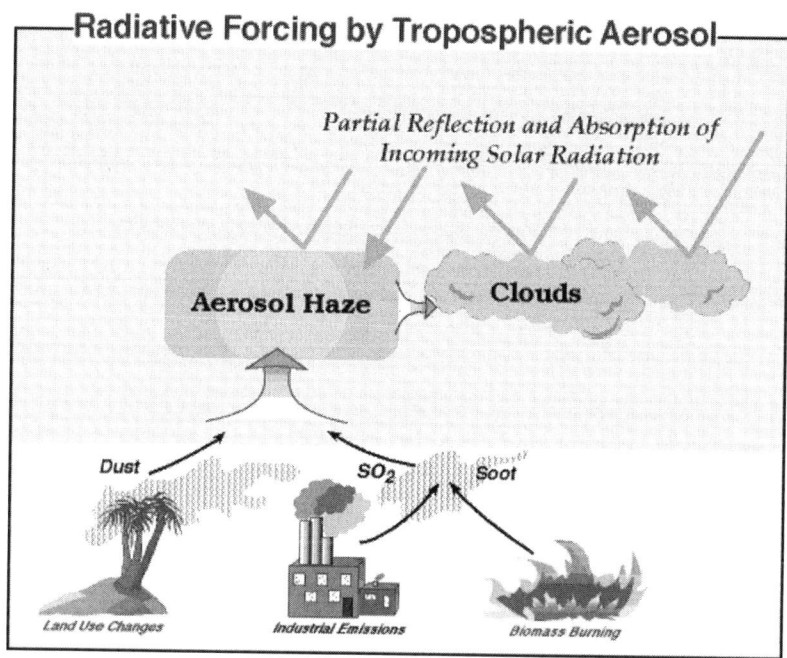

*Aerosols can both reflect and absorb solar radiation, making their role in the greenhouse effect particularly complex.* (NOAA)

Aerosols in industrialized areas often contain heavy metals—including chromium (Cr), iron (Fe), copper (Cu), cadmium (Cd), cobalt (Co), nickel (Ni), and lead (Pb)—latex, surfactants, and asbestos. When tetraethyl lead was used as a gasoline additive, inhalation of lead aerosols contributed a substantial fraction of the body burden of lead in urban dwellers. Recognition of this hazard led to a ban on leaded gasoline. Dry deposition of heavy metals in aerosols can have adverse effects on ecosystems; $CdSO_4$ and $CuSO_4$ are known to reduce root elongation in trees. Fluorides released by heavy industry contribute to tree dieback.

The adverse impact of aerosols on air quality has been noted by many writers since the time of John Evelyn, who, on January 24, 1684, recorded a marked decrease in atmospheric visibility and increased respiratory problems associated with London smog. During the late nineteenth and early twentieth centuries, the lethal effects of aerosols were evident in the greatly increased mortality that occurred during London smog episodes. A historic episode in 1952, during which smog in London reduced visibility to zero, caused an estimated four thousand excess deaths, doubling the normal death rate for children and for adults ages forty-five to sixty-four. This episode led to legislation regarding emissions controls, and by the

late 1960's, the health effects related to coal burning were reduced to minimal levels.

Urban aerosols—with their mix of latex, soot, hydrocarbons, $SO_x$ and $NO_x$, and other pollutants— have been implicated in the growing prevalence of asthma. In the United States, the National Institutes of Health (NIH) estimated that asthma prevalence rose 34 percent between 1983 and 1993. About 4.8 million children in the United States were estimated to suffer from asthma in 1993. Asthma deaths have rapidly increased in the United States, rising from 0.9 per 100,000 people in 1976 to 1.5 per 100,000 people in 1986. Deaths from asthma among African Americans of all ages rose from 1.5 per 100,000 in 1976 to 2.8 per 100,000 in 1986. African Americans between the ages of fifteen and twenty-four had an asthma death rate of 8.2 per one million in 1980; by 1993, the rate had increased to 18.8 per one million.

Aerosols also increase the absorption and scattering of light in the atmosphere, reducing visibility. When smoke particle concentrations exceed 80-100 milligrams per cubic meter ($mg/m^3$), visibility falls below 1 kilometer (0.62 miles). Smoke palls can travel considerable distances; for example, during the spring of 1998, agricultural burning in the Yucatán in Mexico created a pall that markedly diminished visibility in Dallas, Texas.

Particulates in aerosols serve as condensation nuclei and may enhance fog formation. Visibility reductions by hygroscopic (moisture-retaining) air pollutants are noticeable at relative humidities of about 50 percent. A number of multicar chain-reaction accidents have occurred downwind of industrial plants when hygroscopic plumes have passed over cooling ponds adjacent to highways, creating abrupt reductions in visibility. When drivers of cars and trucks moving at high speeds suddenly become engulfed in an area with visibility of only a few feet, they often respond erratically, causing accidents.

Extremely low air temperatures can lead to the formation of ice fogs—known as "arctic haze"—over cities; such fogs, which are entirely anthropogenic (caused by humans), are fed by moisture and by particles given off by combustion. They are composed of minute ice crystals that substantially reduce visibility to the point where air travel is restricted. Ice fogs last for several days at a time and are frequent during winters in Fairbanks, Alaska, and many Canadian cities. The Siberian city of Irkutsk reports an average of 103 days of fog yearly, all during winter.

*Anita Baker-Blocker*

FURTHER READING

Colbeck, Ian, ed. *Environmental Chemistry of Aerosols.* Ames, Iowa: Blackwell, 2008.

Cotton, William R., and Roger A. Pielke, Sr. *Human Impacts on Weather and Climate.* 2d ed. New York: Cambridge University Press, 2007.

Katsouyanni, K., et al. "Short-Term Effects of Ambient Sulphur Dioxide and Particulate Matter on Mortality in Twelve European Cities: Results from Time Series Data from the APHEA Project." *British Medical Journal* 314 (June 7, 1997).

# Air-conditioning

CATEGORY: Atmosphere and air pollution

DEFINITION: Intentional modification of indoor air temperature, humidity, or other characteristics for the purpose of comfort, health, or protection of items sensitive to temperature or humidity

SIGNIFICANCE: Air-conditioning benefits humankind in many ways, but it also has some negative environmental impacts. The refrigerants long used in air conditioners have proven to have adverse effects on ozone in the earth's stratosphere, and the use of air-conditioning accounts for a high percentage of energy consumption, particularly in regions with warm climates.

Since ancient times, human beings have tried to make the air inside buildings more comfortable. The Romans circulated cool water through buildings. Other cultures created fans or encouraged breezes to blow across pools of evaporating water to cool buildings. Cooling by evaporation of water, however, can lower temperatures only by a few degrees and is not effective when the air is humid.

Most modern air conditioners are similar to that first proposed by Willis Carrier in 1902. Air condition-

ers work in much the same way as do electric refrigerators. Both kinds of devices make use of the thermodynamics of fluids. A fluid with a boiling point somewhat below room temperature is selected, and a compressor is used to compress the fluid. Under sufficiently high pressure, the fluid liquefies. This process produces heat, so, in a part of the system outside the area to be cooled, fans blow air over tubes containing the warm liquid to cool it. The liquid is then circulated into the part of the system within the area to be cooled, where it passes through tubes in which the pressure drops, allowing the fluid to evaporate. The evaporating fluid then cools to a temperature near its boiling point. A fan blows air over the cold tubes, cooling the air. Many early air conditioners used fluids such as ammonia, methyl chloride, or even propane. The problem with these materials is that they are health hazards, explosion or fire hazards, or both.

In 1928, Thomas Midgley, Jr., developed chlorofluorocarbons (CFCs), a family of related chemicals that were nontoxic and nonexplosive and that had boiling temperatures and other properties making them suitable for use in air conditioners and refrigerators. It was unknown at the time that these nontoxic CFCs, when released into the atmosphere, break down ozone in the earth's stratosphere. Because ozone in the stratosphere blocks ultraviolet light from the sun, maintenance of the ozone layer is important to life on earth. Without the ozone layer, the earth's surface would receive a lethal dosage of ultraviolet light. Some damage has already been done to stratospheric ozone by CFCs released from leaking air-conditioning systems. In addition to depleting stratospheric ozone, CFCs are greenhouse gases, which means that they contribute to global warming.

The international agreement known as the Montreal Protocol calls for the gradual reduction of the production of CFCs and other chemicals known to harm the ozone layer, and since the adoption of the protocol in 1987 the use of these substances has been greatly reduced around the world. Some non-CFC refrigerants have been developed that are less damaging to the environment; however, these newer refrigerants have slightly different properties and cannot simply be substituted for CFCs in older air-conditioning systems without modifications to the systems, some of which can be expensive.

In regions with warm climates, air-conditioning accounts for the largest use of electricity in buildings. With electricity becoming increasingly expensive, many manufacturers of air conditioners emphasize energy efficiency in their marketing. The standard reference for air-conditioner energy efficiency is the seasonal energy-efficiency ratio (SEER), which is the thermal energy removed (measured in British thermal units) divided by the energy used (measured in watt-hours). The larger the SEER, the more energy-efficient the air conditioner.

*Raymond D. Benge, Jr.*

FURTHER READING

Achermann, Marsha E. *Cool Comfort: America's Romance with Air-Conditioning.* Washington, D.C.: Smithsonian Institution Press, 2002.

Miller, Rex, and Mark R. Miller. *Air Conditioning and Refrigeration.* New York: McGraw-Hill, 2006.

## Percentage Change in U.S. Emissions
(millions of tons per year)

|  | 1980 vs. 2008 |
|---|---|
| Carbon monoxide | −56 |
| Lead | −99 |
| Nitrogen oxides | −40 |
| Volatile organic compounds | −47 |
| Direct particulate matter (10-micron-diameter) | −68 |
| Direct particulate matter (2.5-micron-diameter) | — |
| Sulfur dioxide | −56 |

*Source:* Data from U.S. Environmental Protection Agency, *Air Quality Trends,* 2009.

# Air pollution

CATEGORIES: Atmosphere and air pollution; weather and climate

DEFINITION: Atmospheric presence of materials or energy in quantities sufficient to harm living organisms, affect weather changes, or damage human-made materials and structures

SIGNIFICANCE: Air pollution's effects are profound and far-reaching. Acute pollution episodes can cause human fatalities, while ongoing pollution at lower concentrations results in cumulative damage to health. Pollutants can be transported long distances from their sources, where they can affect populations in neighboring cities, states, or nations. Pollutants such as ozone-depleting substances and greenhouse gases can have global impacts.

Earth's atmosphere is a mixture of two types of gases: The gases of one type are present in concentrations that are constant over long periods of time, and the gases of the other type are present in variable concentrations. Among the former are nitrogen ($N_2$), which makes up approximately 78 percent of the atmosphere, and oxygen ($O_2$), which constitutes about 21 percent of the atmosphere. Along with these two are argon (Ar), almost 1 percent of the atmosphere, and trace amounts of neon (Ne), helium (He), krypton (Kr), hydrogen ($H_2$), and xenon (Xe), none of which seems to have any major effect on the atmosphere. Among the gases that vary in concentration are carbon dioxide ($CO_2$), water vapor ($H_2O$), methane ($CH_4$), carbon monoxide (CO), ozone ($O_3$), ammonia ($NH_3$), hydrogen sulfide ($H_2S$), and several oxides of nitrogen and sulfur. Water vapor has the highest degree of variability and has a significant effect on the atmosphere because of its ability to change phase readily, absorbing or emitting energy as it does so. The trace components of the atmosphere are continually redistributed by the complex circulation patterns of the air known as winds.

In addition to its chemical makeup, the earth's atmosphere is characterized by the various physical effects acting on it or taking place within it. The most important of these are solar radiation and thermal energy. The sun acts as an almost perfect black-body radiator at an effective temperature of 3,316 degrees Celsius (6,000 degrees Fahrenheit). Some of the incoming solar radiation is absorbed by atmospheric gases such as oxygen, ozone, carbon dioxide, and water vapor, allowing about 80 to 85 percent to reach the ground under clear sky conditions. Cloud cover ensures that only about 50 percent of the solar radiation reaches the planet's surface on the average.

This incident solar radiation is absorbed by the atmosphere and the earth's surface and reradiated at longer wavelengths, mostly infrared. The amount of this radiation that reaches space is affected by atmospheric concentrations of carbon dioxide and water, both of which absorb some infrared radiation. This absorption produces the greenhouse effect, an important process that keeps the lower atmosphere at a higher temperature than the upper atmosphere, supporting life on the planet.

The atmosphere is a chemical system that is not in equilibrium, mainly because of the activities of living organisms. The respiration of many organisms produces carbon dioxide, which, by means of photosynthesis, produces oxygen. The huge amount of oxygen in the atmosphere is almost entirely the result of photosynthesis. Methane, the main hydrocarbon in the atmosphere, is produced by microbial degradation of organic matter in marshes, paddy fields, and the digestive systems of animals. Microorganisms that degrade nitrogen compounds in animal urine create ammonia. Forests are a large source of more complex hydrocarbons such as alkanes, alkenes, and esters (which produce the odors of flowers and fruits).

These natural processes have resulted in an atmosphere that has, since the formation of the planet, reached a steady-state composition. Since the nineteenth century's Industrial Revolution, however, human activities have significantly changed the amounts of some of the compounds found naturally in the atmosphere and have introduced many new compounds. Some of these substances are toxic to plant and animal life. In addition, some anthropogenic (human-caused) compounds have the potential to affect seriously the delicate balance that exists among the earth, its atmosphere, and living organisms.

## Acid Rain

The acidity of precipitation is usually expressed in terms of its pH values, where pH represents the concentration of hydrogen ions present. The pH scale commonly in use extends from 0 to 14, with the pH value 7 representing neutral solutions, values greater than 7 basic solutions, and values less than 7 acidic solutions. The pH scale is logarithmic, so that an increase of one pH unit corresponds to a tenfold increase in hydrogen ion concentration. Unpolluted precipitation has a pH value of around 5.6. This natural acidity results mainly from the atmospheric presence of carbon dioxide, which forms carbonic acid, and of chlorine from sea salt, which forms hydrochloric acid. Other natural contributions that affect pH, regionally or globally, are ammonia, soil particles, and volcanic emissions of sulfur dioxide and hydrogen sulfide. Testing of precipitation that predates the Industrial Revolution, which has been preserved in glaciers, generally has found a pH of more than 5, sometimes as high as 6.

Precipitation with a pH value of less than 5.6 is generally referred to as acid rain. Its average pH values commonly range from 4 to 5; as of 2000, the most acidic rainfalls in the United States had a pH of around 4.3. Individual storms may be much more acidic, such as a rainfall in West Virginia in 1978 that had an unofficial pH of 2.0. The major causes of acid rain are human-generated sulfur dioxide and nitrogen dioxide. Ice cores covering the period between 1869 and 1984 have shown that by the mid-1980's precipitation in Greenland had experienced a threefold increase in sulfate ion concentrations since 1900, and nitrate concentrations had undergone a twofold increase since 1955. These increases are believed to be the result of sulfur oxide and nitrogen oxide emissions carried from North American and Eurasian sources.

The fact that acid rain is largely a regional phenomenon gives clues to the sources of the major contributors. Much acid precipitation results from the combustion of fossil fuels, especially high-sulfur coal. The sulfur is oxidized by burning into sulfur dioxide ($SO_2$), which is released into the atmosphere with the main combustion products, carbon dioxide and water. In the atmosphere, sulfur dioxide reacts with water to produce sulfuric acid ($H_2SO_4$). This in turn falls to the earth's surface as both wet and dry acidic deposition. High atmospheric concentrations of sulfur dioxide and soot particles produce a grayish haze known as "London smog," a particularly severe occurrence of which affected London, England, in December, 1952. During a four-day period, many people in London experienced respiratory difficulty, and thousands of deaths from respiratory causes directly paralleled measured average smoke and sulfur dioxide concentrations.

Nitrogen and oxygen in the air do not react at any significant rate but readily combine to form nitrogen oxides ($NO_x$) in the high-temperature combustion processes found in power plants, smelters, steel mills, and internal combustion engines. In the atmosphere, gaseous nitrogen oxides produce the brownish haze often seen over cities such as Los Angeles and Denver. The nitrogen oxides go through various reactions in the atmosphere, some of which result in the production of nitric acid ($HNO_3$), which eventually reaches the earth's surface.

Among the effects of these acids are the corrosion of human-made objects such as metallic structures, stone buildings, statues, and automotive paints and other coatings. Lakes and other surface waters are affected by acid rain not only through direct alteration of their pH but also by metals such as aluminum, manganese, zinc, nickel, lead, mercury, and cadmium, which are leached from surrounding soils by the overly acidified precipitation. These alterations in water chemistry can kill aquatic organisms outright or affect their health and the viability of their offspring. Acid rain also harms trees and other plants, weakening them by damaging their leaves, stripping soils of necessary nutrients, and releasing substances from soils that are toxic to plant life.

In the United States a program to reduce acid rain that began in 1995 has made significant progress. While the problem of acid rain has not been eliminated, by 2009 the program had reduced annual sulfur dioxide emissions from electricity-generating units by 67 percent compared with 1980 levels and 64 percent compared with 1990 levels. From 1995 to 2009 annual emissions of nitrogen oxides from these units dropped by 67 percent.

## OZONE

Ozone is a form of oxygen found in small quantities throughout the earth's atmosphere. In the troposphere (the lowest layer of the atmosphere), ozone is of interest for a number of reasons. First, it plays an important role in the control of photochemistry, a group of processes in which compounds produced in the reduced state from natural or anthropogenic sources are oxidized to chemically inert materials such as carbon dioxide or to materials that can be precipitated from the atmosphere, such as nitric acid. Photochemical reactions in the troposphere provide the chief cleansing mechanism by which some materials are removed from the atmosphere. The importance of ozone to this process arises from its dissociation by ultraviolet radiation to produce reactive atomic oxygen. Some of the atomic oxygen reacts with water to produce hydroxyl **radicals**, which are responsible for the oxidation of most trace gases.

Ozone in the troposphere is also an important pollutant. It is implicated in the breakdown of natural polymers such as rubber, cotton, leather, cellulose, some paints, plastics, nylon, and fabric dyes. Because ozone is a very strong oxidant, it is a potential irritant to the lungs of humans and animals. Finally, tropospheric ozone, because of its oxidizing ability, is involved in global climate control because of its ability to influence concentrations of such greenhouse gases as carbon dioxide and methane. In addition, ozone is itself a greenhouse gas.

In the stratosphere, ozone provides an essential umbrella that partially shields the earth's surface from dangerous ultraviolet radiation. In

---

## Major Air Pollutants

*Pollution of the earth's atmosphere comes from many sources. Some sources are natural, such as volcanoes and lightning-caused forest fires, but most sources of pollution are by-products of industrial society, such as that of Donora, Pennsylvania. Each of the following eight major forms of air pollution has an impact on the atmosphere. Often two or more forms of pollution have a combined impact that exceeds the impact of the two acting separately.*

1. **Suspended particulate matter:** This is a mixture of solid particles and aerosols suspended in the air. These particles can have a harmful impact on human respiratory functions.

2. **Carbon monoxide (CO):** An invisible, colorless gas that is highly poisonous to air-breathing animals.

3. **Nitrogen oxides:** These include several forms of nitrogen-oxygen compounds that are converted to nitric acid in the atmosphere and are a major source of acid deposition.

4. **Sulfur oxides, mainly sulfur dioxide:** This sulfur-oxygen compound is converted to sulfuric acid in the atmosphere and is another source of acid deposition.

5. **Volatile organic compounds (VOCs):** These include such materials as gasoline and organic cleaning solvents, which evaporate and enter the air in a vapor state. VOCs are a major source of ozone formation in the lower atmosphere.

6. **Ozone and other petrochemical oxidants:** Ground-level ozone is highly toxic to animals and plants. Ozone in the upper atmosphere, however, helps to shield living creatures from ultraviolet radiation.

7. **Lead and other heavy metals:** Generated by various industrial processes, lead is harmful to human health even at very low concentrations.

8. **Air toxics and radon:** Examples include cancer-causing agents, such as radioactive materials and asbestos. Radon is a radioactive gas produced by natural processes in the earth.

the upper atmosphere, ozone formation involves oxygen and ultraviolet radiation. The reaction is $O_2 + h\upsilon \rightarrow O + O$, where $h\upsilon$ is ultraviolet energy. The oxygen atoms then react with oxygen molecules to produce ozone according to the reaction $O_2 + O \rightarrow O_3$. This photochemical process by which ozone is produced is balanced by the photochemical process that destroys it: $O_3 + h\upsilon \rightarrow O_2 + O$. Both processes involve the absorption of ultraviolet radiation, and the dynamic chemical equilibrium that exists between them removes a portion of the ultraviolet energy as it travels toward the earth's surface.

Knowledge of the equilibrium chemistry between oxygen and ozone allowed prediction of the equilibrium concentration of ozone in the upper atmosphere. Since the early 1970's, measurements of stratospheric ozone concentrations over the Antarctic continent have pointed to concentrations much lower (sometimes by as much as 50 percent) than expected. Reductions in ozone concentrations have also been seen over the Arctic and at midlatitudes.

The reason for this ozone depletion was ultimately traced to human activity, notably the widespread use of a group of compounds called chlorofluorocarbons (CFCs). These chemicals were especially effective in air-conditioning and refrigeration systems and as blowing agents in plastic-forming processes, solvents in the electronics industry, and propellants in aerosol spray cans. CFCs were initially believed to be free of side effects, and as a result large amounts of them were expelled into the atmosphere. The very inertness that makes CFCs so useful for industrial and consumer applications, however, became a great disadvantage as the compounds made their long journey to the stratosphere.

Once in the stratosphere, CFC molecules absorb ultraviolet radiation and break down, yielding chlorine atoms. These chlorine atoms catalyze the conversion of ozone to oxygen. Because the chlorine atom is a catalyst in the process, it is released to continue its destructive activity for many years to come. Halons—bromofluorocarbon compounds with excellent fire-suppressant properties—are similarly destructive to ozone, as their molecules break down in the presence of ultraviolet radiation to produce bromine atoms that destroy ozone.

The Montreal Protocol, an international environmental agreement that has undergone several amendments since it was adopted in 1987, has phased out CFCs, halons, and other ozone-depleting substances, with related ozone-depleting compounds to be phased out in future. While overall concentrations of these substances are on the decline in the stratosphere, their long residence time there means that ozone levels are unlikely to rebound to pre-1980 levels any sooner than the mid-twenty-first century.

GLOBAL CLIMATE CHANGE

When energy in the form of electromagnetic radiation strikes a molecule, the energy may be reflected, absorbed, or transmitted. Solar energy striking the earth's surface is absorbed and heats the land and water, which in turn radiate energy in the form of infrared radiation back toward space. Eventually an equilibrium state is reached in which the amount of energy absorbed by the earth equals the amount radiated. In the absence of an atmosphere, the equilibrium temperature of the earth would be about −21 degrees Celsius (−5.8 degrees Fahrenheit).

The atmosphere contains gases that transmit ultraviolet and visible radiation but absorb infrared wavelengths. Therefore, the infrared energy radiated by earth toward space is trapped in the air layer, increasing its temperature and that of the earth's surface. The equilibrium temperature of the earth because of this phenomenon is about 12 degrees Celsius (53.6 degrees Fahrenheit), 33 degrees Celsius (59.4 degrees Fahrenheit) warmer than it would be without those gases. These gases do for the earth what glass walls and roofs do for the temperature of a greenhouse; therefore, they are known as greenhouse gases.

Any molecule with two or more atoms that has no center of symmetry is a potential greenhouse gas. Important greenhouse gases in the earth's atmosphere include carbon dioxide, methane, nitrous oxide, ozone, and CFCs. This collection of gases absorbs radiation across the infrared range of wavelengths so that there are no windows for reflected infrared radiation to escape back into space. Since preindustrial times, atmospheric concentrations of greenhouse gases have been rising because of human activity. Between 1970 and 2004, total anthropogenic greenhouse gas emissions increased by approximately 70 percent.

Water vapor is the greatest contributor to the greenhouse effect, but its concentration is generally considered to be unaffected by human activities. After water the most important of the greenhouse gases is carbon dioxide. Carbon dioxide from fossil-fuel use alone accounted for 56.6 percent of the world's anthropogenic greenhouse gas emissions in 2004.

The concentration of carbon dioxide in the atmosphere increased by about 80 percent between 1970 and 2004, primarily from the burning of fossil fuels. Among the products resulting from any hydrocarbon combustion are water and carbon dioxide. All processes that depend on energy from coal, oil, or natural gas are contributing to the total amount of greenhouse gases in the atmosphere.

Whether weather phenomena such as frequent serious storms, El Niño conditions, droughts, and floods are directly related to the greenhouse effect is a topic of heated debate among scientists and nonscientists. Cores covering thousands of years of accumulation taken from the Antarctic ice pack have been examined for clues about concentrations of atmospheric gases and average temperatures. An almost direct correlation has been found between carbon dioxide concentration and surface temperature. The historical evidence seems to point to potentially serious consequences if humankind does not quickly develop and implement the use of forms of energy that do not contribute to carbon dioxide emissions.

OTHER POLLUTANTS

In addition to gases, air contains suspended particulate matter. The particles are collections of molecules, sometimes similar, sometimes different. The constituents of particulate matter differ over time and space. In urban areas, particulate matter often contains sulfuric acid and other sulfates, carbon, or higher molecular weight hydrocarbons that result from incomplete combustion of fossil fuels. Particulate matter and sulfur dioxide are common pollutants found in urban smog. Over time, suspended particles tend to increase mass by combining or acting as nuclei on which vapors condense. Eventually these fall to the ground or are washed out by precipitation.

The greatest concern over particulate matter in the atmosphere is the fact that often the particles are small enough to be inhaled and retained in the respiratory system. Vegetation is affected when particles coat leaves and thus reduce plants' absorption of carbon dioxide and suppress photosynthesis and hence plant growth. Particulate matter adheres to painted surfaces and buildings, reducing the lifetimes of materials and coatings and often causing corrosion, especially in moist atmospheres.

Other pollutants in the atmosphere include radioactive materials, carbon monoxide, lead, and hydrocarbons. Radioactive materials result from natural processes (including the decay of materials such as uranium) and from human nuclear technology. Radioactive nuclides produce ionizing radiation, which has the potential for long-term effects on cells, including cell death, genetic mutations, and malignant tumor formation.

Carbon monoxide results from the incomplete combustion of hydrocarbons. It is an unstable compound that quickly oxidizes to carbon dioxide. It is absorbed through the lungs and forms a complex with hemoglobin that is more tightly bound than oxygen. In this way carbon monoxide prevents oxygen from reaching individual cells, eventually resulting in death.

Metals such as beryllium, cadmium, chromium, lead, manganese, mercury, nickel, and vanadium may also be found in the air. Lead in particular is widely dispersed throughout the environment, mainly because of its use as an additive in gasolines. Most countries have taken major steps to ban leaded gasolines, but residual concentrations still affect humans, especially children in urban areas. Another problematic metal, mercury, can enter the atmosphere through power plant emissions. Many coal deposits contain mercury, which is released when the coal is burned. Airborne mercury can travel far from its source before it settles into water or onto land, where it can enter the food web and bioaccumulate within living organisms.

Hydrocarbons and their derivatives may be found as solids, liquids, or gases. Although some are the results of natural processes, most are by-products of combustion processes. Some of the hydrocarbons are toxic even in small concentrations, but the major contribution of hydrocarbons is their involvement in atmospheric photochemistry.

*Grace A. Banks*
*Updated by Karen N. Kähler*

FURTHER READING

Godish, Thad. *Air Quality.* 4th ed. Boca Raton, Fla.: Lewis, 2004.

Hilgenkamp, Kathryn. "Air." In *Environmental Health: Ecological Perspectives.* Sudbury, Mass.: Jones and Bartlett, 2006.

Hill, Marquita K. "Air Pollution." In *Understanding Environmental Pollution.* 3d ed. New York: Cambridge University Press, 2010.

Jacobson, Mark Z. *Atmospheric Pollution: History, Science, and Regulation.* New York: Cambridge University Press, 2002.

McKinney, Michael L., Robert M. Schoch, and Logan

Yonavjak. "Air Pollution: Local and Regional." In *Environmental Science: Systems and Solutions*. 4th ed. Sudbury, Mass.: Jones and Bartlett, 2007.

Seinfeld, John H., and Spyros N. Pandis. *Atmospheric Chemistry and Physics: From Air Pollution to Climate Change*. 2d ed. Hoboken, N.J.: John Wiley & Sons, 2006.

Sokhi, Ranjeet S., ed. *World Atlas of Atmospheric Pollution*. London: Anthem Press, 2008.

U.S. Environmental Protection Agency. *Our Nation's Air: Status and Trends Through 2008*. Research Triangle Park, N.C.: Author, 2010.

Vallero, Daniel. *Fundamentals of Air Pollution*. 4th ed. Boston: Elsevier, 2008.

# Air-pollution policy

CATEGORY: Atmosphere and air pollution

DEFINITION: High-level governmental plan of action for establishing and maintaining acceptable air quality and regulating individual air pollutants

SIGNIFICANCE: Laws and regulatory agencies establish air-pollution policy to control human-generated pollutants that can have negative impacts on human life and health, ecosystems, and global processes such as stratospheric ozone replenishment. Regulatory policy typically seeks to control pollutants by setting ambient air-quality standards, limiting allowable emissions, and requiring the use of specific pollution-control technologies.

The Clean Air Act, passed by the U.S. Congress in 1963, laid the foundation for what some consider to be the most progressive, wide-reaching, and complicated environmental cleanup legislation in the world. When the Clean Air Act and other early federal, state, and local clean air laws proved to be relatively ineffective, several sweeping amendments to the laws were enacted.

The groundbreaking 1970 amendments to the Clean Air Act resulted in emissions standards for automobiles and new industries in addition to establishing air-quality standards for urban areas. Devised through an exceptionally cooperative bipartisan effort, the 1970 amendments were proclaimed by President Richard M. Nixon to be a "historic piece of legislation" that put the United States "far down the road" toward achieving cleaner air. The amendments established specific maximum concentration levels for several hazardous substances, and the individual states were charged with developing comprehensive plans to implement and maintain these standards.

Tightly controlled scientific methodology was used for the first time to assess and determine acceptable levels for public and environmental health for six "priority air pollutants": carbon monoxide, sulfur dioxide, nitrogen dioxide, respirable particulate matter, ground-level ozone, and lead. Emission standards for air-pollution sources such as automobiles, factories, and power plants were established that also limited the discharge of air pollutants in geographical areas where air quality was already acceptable, thus preventing its deterioration.

The major Clean Air Act amendments of 1970 also stimulated many states to pass regional and local air-pollution legislation, with some areas eventually passing laws that later proved to be even more stringent than federally established guidelines. During this period, the newly created U.S. Environmental Protection Agency (EPA) began strongly suggesting the tightening of rules regulating the amount of lead that could be added to gasoline, a significant source of lead poisoning in urban children and young adults, thus laying the groundwork for the future elimination of all leaded gasolines. Many sectors of the business community challenged the wording of some of the 1970 amendments, arguing that the language was vague and required clarification, particularly regarding the deterioration of air quality in areas that were already meeting federal standards.

## 1977 AMENDMENTS

The 1977 amendments to the Clean Air Act were stimulated by growing public and government awareness of the necessity for further clarification of standards and the increased knowledge that came from a decade of scientific pollution-control research. Industrial areas that were in violation of air-quality standards, called nonattainment areas, were allowed to expand their factories or build new ones only if the new sources achieved the lowest possible emission rates. Additionally, other sources of pollution under the same ownership in the same state were required to comply with pollution-control provisions, and unavoidable emissions had to be offset by pollution reductions by other companies within the same region. These emissions-offset policies forced new industries within geographical regions to make formal requests that existing local companies reduce their pollution production; such situations often resulted in new companies paying the considerable expense of new emissions-control devices for existing companies.

Protection of air quality in regions that were already meeting federal standards sparked congressional debate, as many environmentalists asserted that existing air-quality standards gave some industries a theoretical license to pollute the air up to permitted levels. Rules for the "prevention of significant deterioration" within areas that already met clean air standards were set for sulfur oxides and particulates in 1977, and many individual experts and organizations lobbied for the inclusion of other pollutants, such as ozone, the chief component of smog.

A final major change mandated by the 1977 amendments was the strengthening of the authority of the EPA to enforce laws by allowing the agency to use civil lawsuits in addition to the criminal lawsuits that were previously required. Civil lawsuits have the advantage of not carrying the burden-of-proof requirements needed for criminal convictions; this legal dilemma previously motivated violating companies to take part in lengthy legal battles, as the legal costs were lower than the costs of purchasing and maintaining the necessary pollution-control devices. The EPA was also empowered to levy noncompliance penalties without having to file lawsuits, using the ar-

## Milestones in Air-Pollution Policy

| Year | Event |
| --- | --- |
| 1963 | The Clean Air Act sets aside $95 million to reduce air pollution in the United States. |
| 1970 | The Environmental Protection Agency is established to enforce environmental legislation. |
| 1970 | Clean Air Act amendments establish stricter air-quality standards. |
| 1977 | Additional Clean Air Act amendments extend compliance deadlines established by the 1970 amendments and allow the EPA to bring civil lawsuits against companies that do not meet air-quality standards. |
| 1979 | The United Nations sponsors the Convention on Long-Range Transboundary Air Pollution, which is designed to reduce acid rain and air pollution. |
| 1987 | The Montreal Protocol is signed by twenty-four nations pledging to reduce the output of ozone-depleting chlorofluorocarbons. |
| 1990 | Clean Air Act amendments increase regulations on emissions that cause acid rain and ozone depletion and also establish a system of pollution permits. |
| 1997 | The Environmental Protection Agency issues updated air-quality standards. |
| 1998 | California institutes tougher emission control standards for new cars; other states follow with similar laws. |
| 2003 | Proposed Clear Skies Act is designed to amend the Clean Air Act with a cap-and-trade system. |
| 2005 | EPA's Clean Air Interstate Rule (CAIR) begins a cap-and-trade program to keep air pollution generated in one state from rendering other states noncompliant with air-quality standards. |
| 2008 | A federal appeals court rules that CAIR exceeds the EPA's regulatory authority but later orders temporary reinstatement. |
| 2009 | The EPA officially finds that the greenhouse gases methane, carbon monoxide, nitrous oxide, hydrofluorocarbons, perfluorocarbons, and sulfur hexafluoride constitute a threat to the public health and welfare. |
| 2010 | The American Lung Association reports that about 58 percent of Americans endure unhealthy air-pollution levels. |
| 2010 | The EPA replaces CAIR with the Transport Rule, requiring eastern states to decrease power plant emissions severely by 2014. |

gument that violators have an unfair business advantage over competitors that are currently complying with established legislation. Additionally, several "right-to-know" laws went into effect beginning in 1985 that required manufacturing plant managers to make health and safety information regarding toxic materials available to current and prospective employees, business partners, and sponsors.

## 1990 AMENDMENTS

In 1990 the Clean Air Act was further amended to address inadequacies in previous amendments, with major changes including the establishment of standards and attainment deadlines for 190 toxic chemicals. The amendments were approved through the same kind of bipartisan effort as the one that resulted in the 1970 amendments, prompting President George H. W. Bush to state that the new legislation moved society much closer toward the clean air environment that "every American expects and deserves."

The 1990 amendments established a market-based measure for pollution taxes on toxic chemical emissions, thus enhancing the incentive for businesses to comply as quickly as possible. Emissions standards were tightened for automobiles, and mileage standards for new vehicles were raised; these provisions attacked the pollution problem at its center by prompting numerous significant steps toward improved fuel efficiency. Notable results of these measures included significant reductions in vehicular emissions of sulfur dioxide and nitrogen oxide (50 percent), carbon monoxide (70 percent), and other harmful substances (20 percent).

The 1990 amendments also established market-based incentives to reduce nitrogen and sulfur oxides because of their role in the growing controversy regarding acid deposition within rainwater. The EPA was empowered to create tradable permits that stipulated permissible emissions levels for nitrogen and sulfur oxides. The permits were issued to U.S. companies that had emission rates lower than those set by current requirements for the improvement of air quality. This landmark legislation enabled companies that implemented innovative and cost-effective means to reduce air pollution to sell their unused credits to other companies.

Other significant legislation passed within or assisted by the 1990 amendments included the beginning of the phasing out of numerous ozone-depleting chemicals and the implementation of strategies that

would help sustain the environment. Discovery of a seasonal "ozone hole" over Antarctica in 1985 had sparked international concern regarding the state of the earth's ozone layer and its ability to continue to shield the planet's surface from harmful ultraviolet radiation. Many businesses complained that the considerable additional expenses associated with implementing these new laws created unnecessary burdens for industry that in many cases outweighed the potential environmental benefits. In some cases, this economic pressure merely transferred environmental problems elsewhere, with many businesses choosing to operate outside the United States, in countries with less stringent environmental requirements.

Another important clause in the 1990 amendments required the EPA to regulate emissions coming from solid waste incinerators, including incinerators used for disposal of medical waste. Medical waste incinerators are among the largest sources of airborne dioxin and mercury, which are widely believed to contribute to serious health problems.

## SUBSEQUENT DEVELOPMENTS

President Bill Clinton continued tightening acceptable levels of smog and soot in the United States but did begin allowing flexible methods for reaching these improved goals over a ten-year period. This marked a significant change from the earlier administration of President Ronald Reagan, which proposed a relaxation of environmental standards to favor industrial and technological interests. Clinton is credited with associating the problem of controlling fossil-fuel emissions with the threat of global warming, an issue that would undergo considerable debate within the United Nations and elsewhere for years to come. A 1990 amendment requiring the use of gasoline containing 2 percent oxygen by weight in regions classified as being in severe or extreme nonattainment for the federal ozone standard was followed by several state-level requirements.

The clean air changes of 1990 led the California Air Resources Board (CARB) to introduce the country's most stringent vehicle emissions quality controls to date later that year. Under the state's ambitious program, 2 percent of all new cars sold in California in 1998 were to have pollution-control devices that released no environmentally harmful emissions at all, and the figure was required to rise to 10 percent by 2003. These monumental state laws also dictated that the hydrocarbon emissions of all new cars sold in California be at least 70 percent less than those sold in 1993 by the year 2003. Thirteen northeastern states later passed similar, but somewhat less rigorous, laws; only New York retained the 2 percent goal.

Given 1990's technology, the CARB standards were in effect a mandate for the automobile industry to develop battery-powered electric vehicles. Automakers failed to mass-produce cost-effective, high-performance, battery-powered electric vehicles soon enough to meet regulatory requirements, however. When California adjusted its requirements in response to this slow progress, the result was a lowering of pressure on the auto industry to meet regulatory demands through innovation. By the early years of the twenty-first century, zero-emission vehicles had yet to become an industry standard, but low-emission vehicles such as hybrids had become commonplace, bringing the industry closer to realizing those goals.

During President George W. Bush's administration, the Clear Skies Act of 2003 sought to amend the Clean Air Act with a cap-and-trade system. The controversial bill was never enacted, but measures from it were included in the EPA's 2005 Clean Air Interstate Rule (CAIR), a cap-and-trade program to keep air pollution generated in one state from rendering another state noncompliant with air-quality standards.

In July, 2008, a federal appeals court ruled that CAIR exceeded the EPA's regulatory authority, but five months later the court ordered a temporary reinstatement until the EPA could develop a satisfactory replacement rule. A proposed replacement, known as the Transport Rule, was issued in July, 2010. The Transport Rule is intended to improve air quality in the eastern United States through a decrease in power plant emissions in thirty-one states and the District of Columbia. It requires that, by 2014, power plants reduce their sulfur dioxide emissions by 71 percent and their nitrogen oxides emissions by 52 percent. Each state must meet firm emissions requirements by 2014, which leaves room for only limited trading of pollution credits. The EPA estimates that the rule, if implemented, could prevent 14,000 to 36,000 premature deaths per year, as well as hundreds of thousands of cases of upper respiratory illness.

In late 2009 the EPA issued findings that current and projected atmospheric concentrations of the greenhouse gases (GHGs) methane, carbon monoxide, nitrous oxide, hydrofluorocarbons, perfluorocarbons, and sulfur hexafluoride constitute a threat to the public health and welfare. While this EPA ac-

tion did not impose regulatory requirements, it paved the way for the agency to finalize GHG emissions standards for new motor vehicles, which contribute to atmospheric GHG concentrations. The EPA also proposed GHG emissions thresholds that would define whether Clean Air Act permits for such emissions are required. These thresholds would target the nation's largest stationary sources, such as power plants, refineries, and cement production facilities, but would not affect small businesses and farms. This proposed rule became the focus of a host of legal challenges.

## INDOOR AIR QUALITY

Indoor air pollution began to receive serious public attention in the United States in the wake of the 1970's energy crisis. Interest in conserving energy drove changes in the construction of new buildings, and the retrofitting of old ones, aimed at retaining desired indoor temperatures. Making structures more airtight meant that air contaminants were also retained indoors. By the late 1980's, enough cases had emerged of people experiencing discomfort or various debilitating health problems—chronic respiratory issues, sinus infections, sore throats, headaches, and more—as a result of time spent in particular buildings that the phenomenon had been dubbed sick building syndrome (SBS). The U.S. Occupational Safety and Health Administration has estimated that 30 to 70 million American workers are affected by SBS, although the vast majority do not suffer serious health problems as a result of exposure.

With most Americans spending more than 90 percent of their lives working, learning, and spending leisure time indoors, indoor air quality has the potential to have profound impacts on the population's health. Despite this fact, no comprehensive federal legislation has addressed indoor air quality in the United States; rather, various federal and state regulatory standards address indoor air quality by focusing on specific pollutants, activities, and types of structures.

Common indoor air pollutants include radon (a natural breakdown product of uranium in soil or rock), tobacco smoke, asbestos, formaldehyde, biological contaminants (such as mold and mildew), combustion products (such as carbon monoxide), cleansers and other household products, and pesticides. A number of federal and state laws address asbestos, including the 1976 Toxic Substances Control Act (TSCA), which gives the EPA broad authority to control the production, distribution, and disposal of

## National Ambient Air Quality Standards for Criteria Pollutants

| POLLUTANT | AVERAGING TIME | POLLUTANT LEVEL | EFFECTS ON HEALTH |
|---|---|---|---|
| **Carbon monoxide:** colorless, odorless, tasteless gas; it is primarily the result of incomplete combustion; in urban areas the major sources are motor vehicle emissions and wood burning. | 1-hour<br><br>8-hour | 35 ppm<br><br>9 ppm | The body is deprived of oxygen; central nervous system affected; decreased exercise capacity; headaches; individuals suffering from angina, other cardiovascular disease; those with pulmonary disease, anemic persons, pregnant women and their unborn children are especially susceptible. |
| **Ozone:** highly reactive gas, the main component of smog. | 1-hour<br><br>8-hour | 0.120 ppm<br><br>0.080 ppm | Impaired mechanical function of the lungs; may induce respiratory symptoms in individuals with asthma, emphysema, or reduced lung function; decreased athletic performance; headache; potentially reduced immune system capacity; irritant to mucous membranes of eyes and throat. |
| **Particulate matter < 10 microns (PM10):** tiny particles of solid or semisolid material found in the atmosphere. | 24-hour<br><br>Annual arithmetic mean | $150\,\mu g/m^3$<br><br>$50\,\mu g/m^3$ | Reduced lung function; aggravation of respiratory ailments; long-term risk of increased cancer rates or development of respiratory problems. |
| **Particulate matter < 2.5 microns (PM2.5):** fine particles of solid or semisolid material found in the atmosphere. | 24-hour<br><br>Annual arithmetic mean | $65\,\mu g/m^3$<br><br>$15\,\mu g/m^3$ | Same as PM10 above. |
| **Lead:** attached to inhalable particulate matter; primary source is motor vehicles that burn unleaded gasoline and re-entrainment of contaminated soil. | Calendar quarter | $1.5\,\mu g/m^3$ | Impaired production of hemoglobin; intestinal cramps; peripheral nerve paralysis; anemia; severe fatigue. |
| **Sulfur dioxide:** colorless gas with a pungent odor. | 3-hour<br><br>24-hour<br><br>Annual arithmetic mean | 0.5 ppm<br><br>0.14 ppm<br><br>0.03 ppm | Aggravation of respiratory tract and impairment of pulmonary functions; increased risk of asthma attacks. |
| **Nitrogen dioxide:** gas contributing to photochemical smog production and emitted from combustion sources. | Annual arithmetic mean | 0.053 ppm | Increased respiratory problems; mild symptomatic effects in asthmatics; increased susceptibility to respiratory infections. |

*Notes:* ppm equals parts per million and $\mu g/m^3$ equals micrograms per cubic meter.
*Source:* United States Environmental Protection Agency (EPA); URL http://www.epa.gov.

potentially hazardous chemicals. Under TSCA, federal standards have also been established for the amount of formaldehyde (a chemical present in many adhesives, resins, and solvents) allowable in composite wood-based products

Other legislation, such as the 1976 Consumer Product Safety Act, has granted federal and state authority over consumer products that are potentially dangerous to public health and the environment, with many products that generate indoor air pollution falling under that jurisdiction. For example, carbon monoxide and other hazardous combustion products can be emitted by stoves, and formaldehyde can outgas from plywood and textiles.

Developing countries have different problems related to indoor air pollution than do the United States and other developed nations. In comparison with developed nations, developing nations generally have fewer and less restrictive environmental regulations in place, and indoor air pollution in these nations is less likely to be caused by the airtightness of buildings than by the use of substances indoors that can be harmful. For example, wood, dung, and crop residues are primary sources of cooking and heating fuels in many developing nations, and these can generate unhealthful air pollutants when burned. The United Nations is involved in various efforts to address the problem of indoor air pollution in developing nations.

## AIR QUALITY

In July, 1997, the EPA issued updated air-quality standards following the most complete scientific review process in the history of the organization. Based on the findings of this review, which was conducted by hundreds of internationally recognized scientists, industry experts, and public health officials, major steps were taken toward the improvement of environmental and public health through the revision of ozone standards for the first time in twenty years. In addition, annual exposure standards for fine particulate matter were introduced. (Short-term standards for coarse and fine particulates had been in place for a decade. Short-term standards that applied specifically to fine particulates were not introduced until 2006.) The EPA's 1997 study concluded that many previously imposed standards were not resulting in enough protection for the environment and public health. Data indicated that repeated exposure to pollutants at levels previously considered to be acceptable could cause permanent lung damage in children and in adults who regularly exercise and work outdoors in many urban environments.

The EPA regularly reviews national air-quality standards for the Clean Air Act's six priority air pollutants. Between 1990 and 2008 the Clean Air Act and its supporting legislation enabled national emissions reductions of 78 percent for lead, 14 percent for ozone, 68 percent for carbon monoxide, 35 percent for nitrogen dioxide, 59 percent for sulfur dioxide, and 31 percent for respirable particulates. According to one study, between 1980 and 2000 the reduction in particle pollution alone increased life expectancy in fifty-one cities in the United States by an average of five months. Thanks to control programs for chemical plants, dry cleaners, coke ovens, incinerators, and mobile sources, total emissions of toxic air pollutants decreased by approximately 40 percent between 1990 and 2005. Haze and acid precipitation were also on the decline.

Despite the progress that has been made, air pollution remains a critical environmental risk in the United States. In 2008 thirty-one areas in the United States failed to meet ambient air-quality standards for ozone, eighteen areas failed to meet standards for particulates, and two failed to meet the standard for lead. In 2010 the American Lung Association reported that approximately 58 percent of the nation's population was continuing to experience unhealthy air-pollution levels. According to the U.S. Centers for Disease Control and Prevention, chronic lower respiratory diseases were the fourth leading cause of death in the United States in 2007; the asthma death rate for children under nineteen years old increased by nearly 80 percent between 1980 and 2001.

On the international level, a 1987 United Nations conference held in Canada saw twenty-four nations agree to guidelines established to protect the ozone layer through the Montreal Protocol on Substances That Deplete the Ozone Layer. The Montreal Protocol, which by 2010 had been ratified by 196 countries and amended several times, provides a framework for the phaseout of certain ozone-depleting compounds and includes a mechanism through which developed nations can aid developing countries in making this transition. The United States has also accepted several of the protocols of the 1979 Geneva Convention on Long-Range Transboundary Air Pollution, an international agreement that addresses the impacts of air-pollution migration across political boundaries. These protocols include those concerning reduction strategies for emissions of nitrogen oxides, cadmium, lead, mercury, sulfur, volatile organic compounds, and ammonia.

*Daniel G. Graetzer*
*Updated by Karen N. Kähler*

FURTHER READING

Bailey, Christopher J. *Congress and Air Pollution: Environmental Politics in the US.* New York: Manchester University Press, 1998.

Ferrey, Steven. "Air Quality Regulation." In *Environmental Law: Examples and Explanations.* 5th ed. New York: Aspen, 2010.

Godish, Thad. "Regulation and Public Policy." In *Air Quality.* 4th ed. Boca Raton, Fla.: Lewis, 2004.

Kessel, Anthony. *Air, the Environment, and Public Health.* New York: Cambridge University Press, 2006.

Melnick, R. Shep. *Regulation and the Courts: The Case of the Clean Air Act.* Washington, D.C.: Brookings Institution Press, 1983.

Rushefsky, Mark E. "Environmental Policy: Challenges and Opportunities." In *Public Policy in the United States: At the Dawn of the Twenty-first Century.* 4th ed. Armonk, N.Y.: M. E. Sharpe, 2008.

U.S. Environmental Protection Agency. *The Plain English Guide to the Clean Air Act.* Research Triangle Park, N.C.: Office of Air Quality Planning and Standards, 2007.

air pollutants in a testing area are measured regularly, and, based on this information, the area is assigned a number. This number fits into a color-coded rating system in which different tiers correspond to the severity of the health threat the air quality poses; the higher the number, the more severe the threat.

Although a host of air pollutants have the potential to cause adverse health effects, only a handful of criteria contaminants are generally used in the assessment of basic air quality. Individual criteria differ from place to place, but the most common pollutants monitored are suspended particulate matter, airborne lead, ground-level ozone, nitrogen dioxide, sulfur dioxide, and carbon monoxide. These hazardous compounds are capable of causing severe respiratory irritation, heart and circulatory problems, and other negative health effects.

Monitoring sites are typically limited to cities and towns, where pollution levels and population densities are both high. Different levels of pollution may fall into different tiers based on the stringency of a particular agency's approach to measuring air quality. For example, Hong Kong's Air Pollution Index has come under heavy criticism for its relatively lax standards, as the air quality it rates as safe sometimes contains pollutants at levels several times higher than those considered acceptable by the World Health Organization.

*Daniel J. Connell*

# Air Quality Index

CATEGORY: Atmosphere and air pollution
DEFINITION: The U.S. Environmental Protection Agency's tool for indicating the health risks posed by ambient air quality in given areas at particular times
SIGNIFICANCE: With air pollution a growing problem the world over, the Air Quality Index and equivalent tools in other nations have become an indispensable part of regional and federal governments' efforts to convey information on ambient air quality to the general public.

Nations around the world employ a number of variations on the Air Quality Index (AQI) developed by the U.S. Environmental Protection Agency (EPA), but the basic methodology used to assess air quality is the same. The concentrations of various

# Airborne particulates

CATEGORY: Atmosphere and air pollution
DEFINITION: Tiny particles found in the air
SIGNIFICANCE: Some airborne particulates, such as dust, dirt, soot, and smoke, are large enough to be visible to the naked eye, while other forms are so small that they require electron microscopes for detection. The inhalation of microscopic particles can have serious adverse effects on human respiratory and cardiovascular health.

Airborne particulate matter (PM) represents a complex mixture of organic and inorganic substances and varies in size, composition, and origin. Some particles, known as primary particles, are emitted directly from sources such as construction sites, unpaved roads, fields, smokestacks, and fires. Secondary particles are formed by reactions of gases, such as sulfur dioxide and nitrogen oxides, that are emitted from power plants, industrial plants, and automobiles. Secondary particles make up most of the fine-particle pollution in the United States.

Particle pollution contains microscopic solids or liquid droplets that are small enough to travel deep into the lungs and cause serious health problems. Breathing such pollution can lead to respiratory symptoms such as coughing and difficult breathing; it can decrease lung function and aggravate existing asthma. Also associated with exposure to particle pollution are chronic obstructive pulmonary disease and emphysema, chronic bronchitis, irregular heartbeat, nonfatal heart attacks, and premature death in people with heart or lung disease.

The size of airborne particulates is directly linked to their potential for causing health problems. The U.S. Environmental Protection Agency (EPA) has established air-quality standards concerning two sizes (or fractions) of particles: PM10 and PM2.5. PM10 particles are those with a diameter of 10 micrometers or smaller (10 micrometers is equal to 0.004 inch, or one-seventh the width of a human hair); they include both coarse and fine particles. PM10 particles smaller than 10 micrometers can settle in the bronchi and lungs and cause health problems. PM2.5 particles are 2.5 micrometers in diameter or smaller. Particles in this fraction tend to penetrate further, reaching the gas exchange regions of the lung, and even smaller particles (0.1 micrometer or smaller) may pass through the lungs into the bloodstream and affect other organs, particularly the cardiovascular system. These particles can also adsorb harmful gases or other components (such as iron, carcinogens, or ozone) and release them within lung cells. Particles emitted from modern diesel engines are typically 0.1 micrometer or smaller. PM2.5 inhalation can lead to high plaque deposits in the arteries, causing vascular inflammation and atherosclerosis (hardening of the arteries that reduces elasticity and can lead to heart attacks).

The federal Clean Air Act requires the EPA to review the latest scientific information every five years and promulgate the National Ambient Air Quality Standards for six pollutants, among them PM. U.S. air-quality standards for PM were first established in

---

## What Is Haze?

*The Environmental Protection Agency's Office of Air Quality Planning and Standards defines "haze" as follows:*

Haze is caused when sunlight encounters tiny pollution particles in the air. Some light is absorbed by particles. Other light is scattered away before it reaches an observer. More pollutants mean more absorption and scattering of light, which reduce the clarity and color of what we see. Some types of particles, such as sulfates, scatter more light, particularly during humid conditions.

*Where does haze-forming pollution come from?*

Air pollutants come from a variety of natural and manmade sources. Natural sources can include windblown dust, and soot from wildfires. Manmade sources can include motor vehicles, electric utility and industrial fuel burning, and manufacturing operations. Some haze-causing particles are directly emitted to the air. Others are formed when gases emitted to the air form particles as they are carried many miles from the source of the pollutants.

*What else can these pollutants do to you and the environment?*

Some of the pollutants which form haze have also been linked to serious health problems and environmental damage. Exposure to very small particles in the air have been linked with increased respiratory illness, decreased lung function, and even premature death. In addition, particles such as nitrates and sulfates contribute to acid rain formation which makes lakes, rivers, and streams unsuitable for many fish, and erodes buildings, historical monuments, and paint on cars.

---

1971 and were not significantly revised until 1987, when the EPA changed the indicator of the standards specifically to regulate PM10 levels. Ten years later, the agency set a separate standard for PM2.5 particles based on new research findings regarding their link to serious health problems. The 1997 standards also retained but slightly revised the PM10 standards, which were intended to regulate inhalable coarse particles ranging from 2.5 to 10 micrometers in diameter. The EPA revised the air-quality standards for airborne particle pollution in 2006, lowering the acceptable level of PM2.5 over a 24-hour period from 65 micrograms per cubic meter to 35 micrograms per cubic meter (1 cubic meter is roughly equivalent to 35 cubic feet). It retained the 24-hour PM10 standard of 150 micrograms per cubic meter.

*Bernard Jacobson*

FURTHER READING

Hilgenkamp, Kathryn. "Air." In *Environmental Health: Ecological Perspectives*. Sudbury, Mass.: Jones and Bartlett, 2006.

National Research Council. *Research Priorities for Airborne Particulate Matter, IV: Continuing Research Progress*. Washington, D.C.: National Academies Press, 2004.

Peters, Annette, and C. Arden Pope III. "Cardiopulmonary Mortality and Air Pollution." *The Lancet* 360 (October 19, 2002): 1184-1185.

# Asbestos

CATEGORIES: Pollutants and toxins; human health and the environment

DEFINITION: Industrial term for certain silicate minerals that occur in the form of long, thin fibers

SIGNIFICANCE: The adverse health effects of breathing high concentrations of asbestos over prolonged periods have been known since the early 1970's. The federal Clean Air Act of 1963 classified asbestos as a carcinogenic material, and in 1990 the U.S. Environmental Protection Agency established a broad ban on the manufacture, processing, importation, and distribution of asbestos products.

Asbestos-form minerals are natural substances that are common in many types of igneous and metamorphic rocks found over large areas of the earth. Erosion continually releases these fibers into the environment, and most people typically inhale thousands of fibers each day, or more than 100 million over a lifetime. Asbestos fibers also enter the body through drinking water. Drinking-water supplies in the United States typically contain almost 1 million fibers per quart, but water in some areas may have as many as 100 million or more fibers per quart.

## PROPERTIES

Many silicate minerals occur in fibrous form, but only six have been commercially produced as asbestos. In order of decreasing commercial importance, these are chrysotile (white asbestos), crocidolite (blue asbestos), amosite (brown asbestos), anthophyllite, tremolite, and actinolite. All these minerals except chrysotile are members of the amphibole group of minerals, which have a chainlike arrange-

ment of atoms. In contrast, chrysotile, as a member of the serpentine family, has atoms arranged in a sheetlike fashion.

Although the individual properties of these minerals differ greatly from one another, they share several characteristics that make them useful and cost-effective. These include great resistance to heat, flame, and acid attack; high tensile strength and flexibility; low electrical conductivity; resistance to friction; and a fibrous form, which allows them to be used for the manufacture of protective clothing. Asbestos thus was widely used until the 1970's in a great variety of building and industrial products. Such common materials as vinyl floor tiles, appliance insulation, patching and joint compounds, automobile brake pads, hair dryers, and ironing board covers all might have contained asbestos. Most such products now contain one or more of several substitutes for asbestos instead of asbestos itself. However, many of the substitutes may not be hazard-free, a fact that is starting to be recognized by legislators. For example, in 1993 the World Health Organization (WHO) stated that all substitute fibers must be tested to determine their carcinogenicity. Germany now classifies glass, rock, and mineral wools as probable carcinogens.

## HEALTH EFFECTS

The U.S. Department of Health and Human Services classifies asbestos as a carcinogen. Studies leading to this determination were mostly based on asbestos workers who had been exposed to extremely high levels of fibers for many years. These studies concluded that the asbestos workers had increased chances of developing two types of cancer: mesothelioma (a cancer of the thin membrane surrounding the lungs) and cancer of the lung tissue itself. These workers were also at increased risk of developing asbestosis, an accumulation of scarlike tissue in the lungs that can cause great difficulty in breathing and permanent disability. None of these diseases develops immediately; all have long latency periods, typically fifteen to forty years. Contrary to common misconception, exposure to asbestos does not cause muscle soreness, headaches, or any other immediate symptoms. The effects of asbestos exposure typically are not noticed for many years.

It is generally agreed that the risk of developing disease after asbestos exposure depends on the number of fibers in the person's body, how long the fibers have been in the body, and whether or not the person is a

## U.S. End Uses of Asbestos, 1977 vs. 2003

| END USE | METRIC TONS | |
|---|---|---|
| | 1977 | 2003 |
| Cement pipe | 145,000 | — |
| Cement sheet | 139,500 | — |
| Coatings and compounds | 32,500 | 1,170 |
| Flooring products | 140,000 | — |
| Friction products | 83,100 | — |
| Insulation: electrical | 3,360 | — |
| Insulation: thermal | 15,000 | — |
| Packing and gaskets | 25,100 | — |
| Paper products | 22,100 | — |
| Plastics | 7,260 | — |
| Roofing products | 57,500 | 2,800 |
| Textiles | 8,800 | — |
| Other | 30,200 | 677 |

*Source:* Data from the U.S. Geological Survey.
*Note:* U.S. mining of asbestos ended in 2002.

smoker, since smoking greatly increases the risk of developing disease. There is no agreement on the risks associated with low-level, nonoccupational exposure. The U.S. Environmental Protection Agency (EPA) has concluded that there is no safe level of exposure to asbestos fibers, but the Occupational Safety and Health Administration (OSHA) allows up to 0.1 fiber per cubic centimeter of air during an eight-hour workday.

### OTHER CONTROVERSIES

Another area of controversy stems from scientific studies showing that all forms of asbestos are not equally dangerous. Evidence has shown that the amphibole forms of asbestos, and particularly crocidolite, are hazardous, but the serpentine mineral chrysotile—accounting for 95 percent of all asbestos used in the past and 99 percent of current production—is not. For example, one case study involved a school that was located next to a 150,000-ton rock dump containing chrysotile. Thousands of children played on the rocks over a one-hundred-year period, but not a single case of asbestos-related disease developed in any of the children. The difference seems to be in how the human body responds to amphibole

compared to chrysotile. The immune system can eliminate chrysotile fibers much more readily than amphibole, and there is also evidence that chrysotile in the lungs dissolves and is excreted. This remains a controversial area, and the U.S. government still treats all forms of asbestos the same. This is not true of some European governments.

The risk of developing any type of disease from exposure to normal levels of asbestos fibers in outdoor air or the air in closed buildings is extremely low. The calculations of Melvin Benarde in *Asbestos: The Hazardous Fiber* (1990) show that the risk of dying from nonoccupational exposure to asbestos is one-third the risk of being killed by lightning. The Health Effects Institute made similar calculations in 1991 and found that the risk of dying from asbestos is less than 1 percent the risk of dying from exposure to secondary tobacco smoke.

*Gene D. Robinson*

FURTHER READING

Bartrip, Peter. *Beyond the Factory Gates: Asbestos and Health in Twentieth Century America.* New York: Continuum, 2006.

Carroll, Stephen, et al. *Asbestos Litigation.* Santa Monica, Calif.: RAND, 2005.

Castleman, Barry. *Asbestos: Medical and Legal Aspects.* 5th ed. New York: Aspen, 2005.

Chatterjee, Kaulir Kisor. "Asbestos." In *Uses of Industrial Minerals, Rocks, and Freshwater.* New York: Nova Science, 2009.

Craighead, John E., and Allen R. Gibbs, ed. *Asbestos and Its Diseases.* New York: Oxford University Press, 2008.

Deffeyes, Kenneth S. "Asbestos." In *Nanoscale: Visualizing an Invisible World.* Illustrations by Stephen E. Deffeyes. Cambridge, Mass.: MIT Press, 2009.

Dodson, Ronald, and Samuel Hammar, eds. *Asbestos: Assessment, Epidemiology, and Health Effects.* Boca Raton, Fla.: CRC Press, 2005.

McCulloch, Jock, and Geoffrey Tweedale. *Defending the Indefensible: The Global Asbestos Industry and Its Fight for Survival.* New York: Oxford University Press, 2008.

Maines, Rachel. *Asbestos and Fire: Technological Trade-Offs and the Body at Risk.* New Brunswick, N.J.: Rutgers University Press, 2005.

# Atmospheric inversions

CATEGORIES: Atmosphere and air pollution; weather and climate

DEFINITION: Vertical temperature profiles in which air temperature increases with height in the atmosphere

SIGNIFICANCE: Temperature inversions play an important role in trapping anthropogenic (human-caused) pollutants near the earth's surface, leading to the formation of smog and reduced air quality in many metropolitan areas. Temperature inversions also play an important role in the formation of severe thunderstorms and mixed precipitation (such as freezing rain and sleet).

An atmospheric inversion is defined in meteorology as a scenario in which air temperature increases with height. To understand its importance it is necessary to consider the concept in light of the broader structure of the atmosphere. The vertical structure of the atmosphere is characterized by four broad regions; in order of increasing altitude, these are the troposphere, stratosphere, mesosphere, and thermosphere. On average, temperatures decrease with height in the troposphere and mesosphere, and increase with height in the stratosphere and thermosphere. The atmosphere's vertical temperature profile is important because it affects the ability of air to move vertically (that is, to rise and fall) in the atmosphere. Vertical motions are generally permitted in regions where air temperatures decrease with height, and they are suppressed in regions where air temperatures increase with height. The latter condition, known as a temperature inversion, creates a layer in the atmosphere that has the ability to limit, or cap, vertical mixing. The stratospheric inversion is a prime example; it caps vertical motions associated with storms in the troposphere.

Smaller-scale temperature inversions that occur within the troposphere may have significant impacts on air quality at urban and regional levels because they trap polluted air masses close to the ground. A common example of this is a radiation inversion, which often develops during winter months on days with limited wind and low humidity. These conditions allow the ground to cool quickly, thereby causing the air immediately above the ground to cool more than the air aloft. The warmer air aloft forms a stable layer, or cap, over the atmosphere below it. Radiation inversions are an air-quality concern because they often set up during the evening traffic rush hour and trap associated pollutants close to the surface. In general, radiation inversions affect small geographic regions and dissipate during the early morning, when solar radiation raises temperatures near the surface.

A larger-scale subsidence inversion forms when a high-pressure system causes air to subside and warm adiabatically (without gaining or losing heat) over a broad region. Different rates of subsidence between air near the surface and air aloft allow the air aloft to warm to a greater degree, setting up a large-scale temperature inversion. In contrast to radiation inversions, subsidence inversions have the potential to persist for long periods of time. This scenario is common in many metropolitan areas that are known for poor air quality, particularly Los Angeles, California. In Los Angeles, the impact of temperature inversions is magnified by the fact that the city is surrounded on three

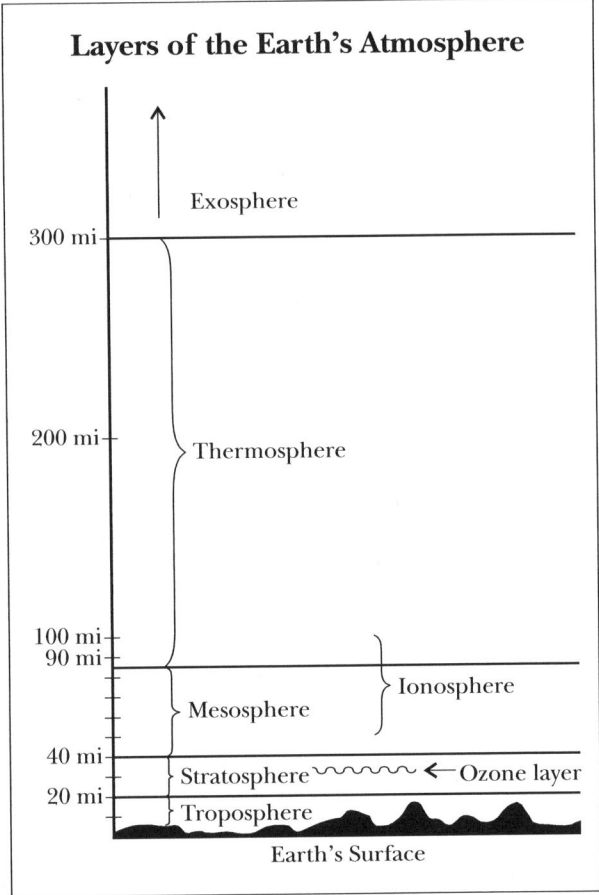

**Layers of the Earth's Atmosphere**

sides by mountains, which limit horizontal mixing and promote higher concentrations of air pollutants in the region.

The pollutants that build up in urban areas as the result of temperature inversions have important long-term health implications and, at times, can prove to be serious short-term threats. For example, in 1948 toxic air conditions that resulted from subsidence inversion in the western Pennsylvania town of Donora caused the deaths of twenty residents. A similar incident known as the London smog disaster killed nearly four thousand residents of London, England, in 1952.

*Jeffrey C. Brunskill*

## FURTHER READING

Aguado, Edward, and James E. Burt. *Understanding Weather and Climate.* 5th ed. Upper Saddle River, N.J.: Pearson Education, 2010.

Ahrens, C. Donald. *Meteorology Today: An Introduction to Weather, Climate, and the Environment.* Belmont, Calif.: Brooks/Cole Cengage Learning, 2009.

# Automobile emissions

CATEGORIES: Atmosphere and air pollution; pollutants and toxins

DEFINITION: Combination of vehicular exhaust from incomplete combustion of gasoline or diesel fuel and hydrocarbons that escape these fuels through evaporation

SIGNIFICANCE: Collectively, the emissions produced by the millions of vehicles on the roads of the world's major metropolitan areas have serious impacts on air quality and human health. They also represent a significant contribution to greenhouse gas emissions. The U.S. Environmental Protection Agency has described driving a private car as "probably a typical citizen's most 'polluting' daily activity."

Automobile emissions create ongoing and potentially dangerous environmental problems when gases and particulates are released into the atmosphere at a rate that exceeds the capacity of the atmosphere to dissipate or dispose of them. Motor vehicle emissions are a major component of the smog that blankets such urban areas as Los Angeles, California, and Denver, Colorado. The emissions produced by a motor vehicle consist of exhaust (the by-products of incomplete gasoline or diesel fuel combustion) and fuel that evaporates from the vehicle's fuel tank, engine, and exhaust system during operation, cooldown, and fueling.

Vehicle exhaust contains several problematic compounds. Carbon monoxide, produced by incomplete combustion, reduces the flow of oxygen in the bloodstream. Hydrocarbons, another product of imperfect combustion, are often toxic or carcinogenic. Nitrogen oxides, which are formed when combusting fuel reacts with oxygen in the air, contribute to the formation of acid rain and fine particles that can harm the lungs when breathed in. Together, hydrocarbons and nitrogen oxides react with the heat of sunlight to produce a hazy brown mixture of secondary pollutants. Notable among these pollutants is ground-level ozone, which irritates the eyes and causes damage to the respiratory system. (Nitrogen oxides also readily create ozone by reacting with naturally occurring hydrocarbons produced by trees.) Other secondary pollutants formed by photochemical reactions include the toxic compounds formaldehyde and peroxyacetyl nitrate. Carbon dioxide, a product of complete combustion, is not directly hazardous to human health, but it is chief among the anthropogenic (human-generated) greenhouse gases commonly believed to affect the world's climate.

## IMPACTS ON HEALTH AND ENVIRONMENT

Environmental problems associated with automobile emissions include deleterious effects on many forms of agriculture and natural forests, reduction in visibility, and damage to metals, building materials such as stone and concrete, rubber, paint, textiles, and plastics. Automobile emissions cause lung and eye irritation, coughing, chest pain, shallow breathing, and headaches. Automobile-produced air pollution is also a factor contributing to allergies, asthma, emphysema, bronchitis, lung cancer, heart disease, and negative psychological states. Carbon monoxide quickly combines with blood hemoglobin and impairs oxygen delivery to the tissues, particularly in children and the elderly, causing heart and lung problems. Vehicular emissions and other sources of air pollution cost Americans billions of dollars each year in health care and related expenses.

The increased rate and depth of breathing during physical exertion exposes delicate lung tissues to more polluted air. Research indicates that exercise

## Costs of Rural and Urban Air Quality Degradation by Motor Vehicles, 2000

| POLLUTANT | IMPACT | RURAL EMISSIONS ($) | URBAN EMISSIONS ($) |
|---|---|---|---|
| Particulate matter | Mortality | 12,695 | 21,558 |
| Particulate matter | Nonfatal illness | 3,683 | 6,232 |
| Sulfur dioxide, nitrogen dioxide, carbon monoxide | Nonfatal illness | 0 | 51 |
| Ozone | Nonfatal illness | 28 | 16 |
| **Total** | | **16,406** | **27,857** |

*Source:* Federal Highway Administration, United States Department of Transportation.

*Note:* Costs of human illnesses, in millions of 1990 dollars. Costs of crop damage, reduced visibility, and other physical effects on the environment are not included.

near a busy freeway may be more harmful than beneficial to the body. At the 1984 Summer Olympics in Los Angeles, the evening rush-hour start of the men's marathon coincided with a stage 2 California health advisory alert, drawing criticism that the organizers of the event were more interested in commercial revenues than in the safety of the athletes and spectators. Later Olympic events were postponed during heavy air-pollution episodes. China faced similar, but more severe, challenges when heavily polluted Beijing hosted the 2008 Summer Olympics. Emergency measures that the city took to improve air quality in time for the games included allowing drivers to use their motor vehicles only every other day, which effectively took more than 1.5 million vehicles off the road daily.

Automobile emissions have been shown to exert their negative effects a considerable distance from the source, depending on atmospheric changes in wind and temperature. Suburbs often exhibit higher levels of pollution than the downtown areas where the emissions are produced. Remote national parks and wilderness areas have had their scenic vistas obscured by haze from distant cities. Fallout of tetraethyl lead from urban automobiles running on leaded gasoline has been observed in oceans and on the Greenland ice sheet. Automobile emissions may also have an impact on a global scale, as they are a significant source of greenhouse gases. Transportation contributed some 13.1 percent of the world's total anthropogenic greenhouse gas emissions in 2004.

### EFFORTS TO REDUCE EMISSIONS

Studies in cities such as London, England, have shown that major improvements in air quality can be achieved in less than ten years in urban areas with favorable climatic conditions through the use of more combustion-efficient engines and cleaner-burning fuels . In the United States, the 1970 amendments to the 1963 Clean Air Act (CAA) introduced automobile emissions standards for hydrocarbons, carbon oxides, and nitrogen oxides and ambient air-quality standards for six pollutants—carbon monoxide, sulfur oxides, nitrogen oxides, particulates, ozone, and lead—to protect human health and the environment. Through the CAA and its amendments of 1970, 1977, and 1990, the Environmental Protection Agency (EPA) has established increasingly stringent emissions-control policies for motor vehicles.

Although the EPA sets pollution standards for vehicles, vehicle manufacturers determine how they will meet those standards. Improved engine design, recirculation of exhaust gas to reduce nitrogen oxides, improved evaporative emissions controls, and computerized diagnostic systems have all led to a decline in polluting emissions. One of the most important milestones in the reduction of hydrocarbon and carbon monoxide emissions was the advent of the catalytic converter in 1975. Because lead impedes the catalyst that reduces emissions, unleaded gasoline became widely available at the same time. Ultimately, leaded gasoline was phased out in the United States and in many other countries, with the result that lead concentrations in ambient air have been lowered significantly.

Although the vehicle emissions controls implemented in the United States since 1970 have been effective, much of their success has been offset by the increasing numbers of vehicles on the roads and the greater distances driven, as many Americans travel farther from their homes to reach workplaces, schools, and shopping and recreation centers. Motor vehicle use has also become more widespread in developing countries, where air-quality standards are often more lax than in developed nations.

In the United States, the pollutant associated with vehicle emissions that has been reduced the most has been lead. Ambient concentrations of carbon monoxide and nitrogen oxides have also decreased; however, while they are low in relation to national standards, they remain a matter of concern because of the role they play in producing ozone and particulates and in impairing visibility. Ground-level ozone, toxic hydrocarbons, and particulates continue to be problems—as does carbon dioxide. In May, 2009, the EPA and the U.S. Department of Transportation agreed to establish national standards for greenhouse gas emissions and fuel economy for new cars and trucks sold in the United States.

*Daniel G. Graetzer*
*Updated by Karen N. Kähler*

FURTHER READING

Godish, Thad. *Air Quality*. 4th ed. Boca Raton, Fla.: Lewis, 2004.

Griffin, Roger D. *Principles of Air Quality Management*. 2d ed. Boca Raton, Fla.: CRC Press, 2007.

Hilgenkamp, Kathryn. "Air." In *Environmental Health: Ecological Perspectives*. Sudbury, Mass.: Jones and Bartlett, 2006.

Jacobson, Mark Z. *Atmospheric Pollution: History, Science, and Regulation*. New York: Cambridge University Press, 2002.

McCarthy, Tom. *Auto Mania: Cars, Consumers, and the Environment*. New Haven, Conn.: Yale University Press, 2007.

Rajan, Sudhir Chella. *The Enigma of Automobility: Democratic Politics and Pollution Control*. Pittsburgh: University of Pittsburgh Press, 1996.

U.S. Environmental Protection Agency. *The Plain English Guide to the Clean Air Act*. Research Triangle Park, N.C.: Office of Air Quality Planning and Standards, 2007.

# Bhopal disaster

CATEGORIES: Disasters; pollutants and toxins

THE EVENT: Release of a highly toxic gas from a pesticide production plant that killed thousands of people and impaired the health of hundreds of thousands

DATES: December 2-3, 1984

SIGNIFICANCE: The escape of methyl isocyanate from a pesticide production plant in Bhopal, India, resulted in the world's worst chemical disaster. The abandoned plant site, which was not properly remediated, continues to pose an environmental threat to nearby residents by contaminating their drinking-water source.

Union Carbide India, which was jointly owned by Union Carbide and the Indian public, ran a pesticide production plant in Bhopal, a city in central India with a population of approximately one million people. A heavily populated shantytown surrounded the plant. Methyl isocyanate (MIC), a very reactive and toxic chemical used in the production of pesticides, was stored at the plant.

As a cost-cutting measure the facility was understaffed, and the plant workers were poorly trained. The plant's signage and manuals were in English, but many of the workers were not English speakers. In 1982, U.S. engineers who conducted a safety audit of the plant noted sixty-one hazards in the unkempt facility. Thirty were deemed critical hazards; of these, eleven were found in the MIC/phosgene units. The audit warned that a major toxic release could result.

On the evening of December 2, 1984, the plant was closed for inventory reduction and routine maintenance. Around 11:30 P.M., workers realized when their eyes started to tear and burn that an MIC leak had occurred. They reported the leak to their supervisor and located a section of open piping that they believed to be the source. They took action to contain this presumed source, and the supervisor went on break. By 12:15 A.M., the pressure and temperature in the MIC storage tank had risen to dangerous levels.

When MIC comes into contact with water, a spontaneous reaction results, releasing heat. In the presence of a variety of catalysts, including iron ions, three molecules of MIC will join together to form a trimer. This reaction also releases heat. It is thought that both of these reactions occurred in the MIC storage tank. The heat released during the chemical reactions raised the temperature of the MIC, which increased the rates of the MIC reactions, releasing even more heat. MIC has a low boiling point, and this heat of reaction caused the MIC to vaporize into a gas. The gas expanded, increasing the pressure inside the storage tank until it burst a rupture disk on the line leading to the pressure release valve. When that safety valve was forced open, the gases from the tank began to escape and formed a lethal cloud that moved across Bhopal.

Some people were killed in their sleep by the toxic gas, while others awoke gasping for breath with their eyes and throats burning. The gas cloud quickly spread over an area of 65 square kilometers (25 square miles), engulfing panicked residents who were trying to flee. In a matter of minutes people began to collapse and die. Pregnant women caught in the cloud spontaneously miscarried.

According to the Indian Council of Medical Research, roughly three thousand people were initially killed, and another fifteen thousand later died from chemical-related illnesses. Approximately fifty thousand would suffer permanent disability. All told, the release affected the health of some half a million survivors. Their children would represent a second generation of victims, as many would be born with birth defects.

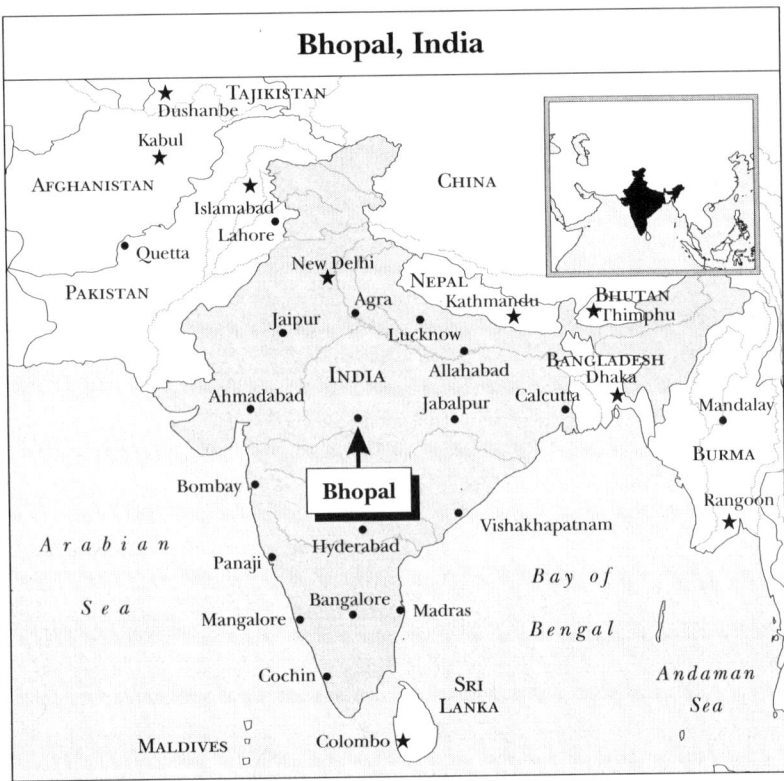

As a result of the Bhopal disaster, India significantly strengthened its regulation of hazardous industries. One of the most important actions the government took was the passage of its Environmental Protection Act of 1986.

### A SERIES OF FAILED SAFEGUARDS

Culpability for the toxic release was found to extend from high levels of management at Union Carbide to workers at the plant. Four months after the disaster, Union Carbide released a report detailing failings that led to the MIC escape. Warren Anderson, the chairman of Union Carbide, acknowledged that conditions at the plant were so poor that it should not have been in operation. The report recounted that sometime before midnight on December 2, 1984, a considerable quantity of water entered the tank and started the heat-releasing reactions. Union Carbide claimed that a disgruntled employee had intentionally introduced the water. Plant workers said it occurred accidentally when water being used to clean pipes leaked into the tank. The tank was equipped with a refrigeration system designed to keep the contents at a low temperature, but it had been shut down months before to cut costs. At the higher storage temperature, the MIC reacted more rapidly with the water.

An alarm that should have sounded when the tank temperature started to rise failed to go off because it had not been reset for the higher storage temperature resulting from the lack of refrigeration. Union Carbide officials hypothesized that the water and high temperature in the tank caused rapid corrosion of its stainless-steel walls. This led to iron contamination that would have catalyzed the heat-releasing trimerization reaction. The temperature in the tank probably rose to at least 200 degrees Celsius (392 degrees Fahrenheit). A control-room operator manually started a sodium hydroxide vent scrubber designed to neutralize leaking MIC, but the sodium hydroxide failed to circulate. The gas rushed to the flare tower, where any MIC escaping the scrubber should have been burned off. This was the last safety control, but it was also out of service. A basic design fault of the plant was that the safety systems were designed to deal with only minor leaks, so even if they had been operative, the volume of gas released would have overwhelmed them.

Although the release started about 12:30 A.M., the

plant's public alarm was not sounded until 2:00 A.M. So many small leaks had triggered it in previous months that the plant had shut it off to avoid disrupting the neighborhood. The local police were informed of the leak only after it had been stopped and after the alarm had been sounded.

## AN ONGOING TRAGEDY

The Indian government sued Union Carbide for $3.3 billion but accepted a settlement in February, 1989, that awarded $470 million to the victims. The next of kin of those killed were each awarded $2,000. Each injured survivor received roughly $500, an amount quickly consumed by medical costs. The tens of thousands who could not navigate the process of filing settlement claims or who were too ill to stand in line for hours to register their claims got nothing. The settlement did not address environmental damages.

A lack of clear, continuing corporate liability, combined with negligence on the part of the Indian government, exacerbated the plight of Bhopal residents. Union Carbide and another company that took over the facility in 1994 performed some cleanup but did not fully remediate the 4.5-hectare (11-acre) site. When the state government took over the property in 1998, hundreds of tons of abandoned toxics remained on-site. In addition to toxic wastes dumped at and around the plant, hundreds of tons of pesticides and other chemicals had been left behind in the abandoned facility. The derelict buildings provided insufficient protection against the elements, allowing rainwater to mix with the chemicals and wash them into the soil. Eventually, the contaminants migrated into the groundwater and thus into the water supplies on which residents relied for drinking and washing needs.

In 2001 Dow Chemical acquired Union Carbide. Because Dow had not operated the plant or caused the accident, it would not acknowledge any responsibility to the affected population. Bhopal residents insisted that Dow had purchased not only Union Carbide's assets but also its liabilities, and they remained unmoved by the contention that Union Carbide's liability ended with the 1989 settlement.

Bhopal citizens have continued to demand that the American-based Dow remediate the site and address the problems that the abandoned facility has created: poisoned cattle, damaged crops, and an array of human health problems including cancers, neurological disorders, mental illness, and birth defects. The victims have also continued to demand action from their government, which they believe has placed greater priority on India's economic advancement and positive relations with multinational corporations than on justice. In 2009 Bhopal victims' groups commemorated the twenty-fifth anniversary of the lethal gas release with vigils and demonstrations.

*Francis P. Mac Kay*
*Updated by Karen N. Kähler*

## FURTHER READING

Amnesty International. *Clouds of Injustice: Bhopal Disaster 20 Years On.* London: Author, 2004.

D'Silva, Themistocles. *The Black Box of Bhopal: A Closer Look at the World's Deadliest Industrial Disaster.* Victoria, B.C.: Trafford, 2006.

Fortun, Kim. *Advocacy After Bhopal: Environmentalism, Disaster, New Global Orders.* Chicago: University of Chicago Press, 2001.

Hanna, Bridget, Ward Morehouse, and Satinath Sarangi, eds. *The Bhopal Reader: Remembering Twenty Years of the World's Worst Industrial Disaster.* New York: Apex Press, 2005.

# Bicycles

CATEGORY: Atmosphere and air pollution

DEFINITION: Pedal-driven vehicles consisting of light frames mounted on two wheels, one in front of the rider and the other behind, steered by means of handlebars

SIGNIFICANCE: The bicycle is the most efficient form of human-powered transportation and is virtually nonpolluting. Many environmentalists advocate the increased use of bicycles for daily transportation in the United States and other developed nations where bicycle use tends to be confined to recreation.

German inventor Karl Drais produced the precursor of the modern bicycle in 1817. The device, known as the *Laufmaschine* (running machine), had a frame that the rider straddled between two wheels of the same size, similar in appearance to today's bicycle. The *Laufmaschine* had no pedals—it was propelled by the rider's feet pushing against the ground. In 1865, the velocipede appeared; this improved on the earlier device with pedals attached to the front wheel. Shortly thereafter, in 1870, the high-wheel bicycle appeared. Also known as the penny-farthing, it had a large front wheel and a much smaller rear wheel (the name penny-farthing was inspired by the relative sizes of the old British penny and farthing coins). The large front wheel allowed faster cycling speeds than were possible on the velocipede, but the high front wheel made it difficult to mount and ride.

By the end of the nineteenth century, the bicycle had evolved to a form that was similar in appearance to the modern bicycle. This vehicle, dubbed the safety bicycle, had two equal-sized wheels with pedals mounted midframe that were connected by a chain to the rear wheel. The safety bicycle sported pneumatic (inflatable) tires, which replaced the hard-riding, solid-rubber tires used on the high-wheel bicycle.

## MODERN BICYCLES

The varieties of bicycles available in the twenty-first century range from basic cruisers to lightweight racing models. An example of a basic cruiser is the single-speed Flying Pigeon, which is widely used as transportation in China. Basic cruisers may have three or more gears. Beyond the basic cruiser are the road bike, designed for use on paved roads, and the mountain bike, intended to be used off-road on hilly terrain. Both types have one, two, or three front chain rings connected to the pedals. Power from the front chain ring is transmitted through a chain to the rear-wheel derailleur, which may have up to ten different gears (cogs); thus bicycles may have up to thirty different gear ratios. Low ratios are used for hill climbing, and high ratios allow for pedaling at high speeds. Both road and mountain bikes come in tandem models, which accommodate two riders. A variation on traditional models is the recumbent bicycle; to use a recumbent bike, the rider sits low to the ground on a seat mounted on the frame between the two wheels rather than straddling a saddle mounted on the frame.

Bicycle riders, or cyclists, must overcome two primary forces: gravity and air resistance. Cyclists have an advantage over runners and walkers in overcoming gravity—gearing aids them in climbing hills, and the downhills give them a chance to recharge. Weight is an important factor in hill climbing—a relatively small rider on a lightweight bicycle has a definite advantage. In regard to air resistance, a stiff headwind can make forward progress difficult for cyclists even on a level surface. In the absence of wind, air resistance increases exponentially with speed. Crouching on a bicycle decreases wind resistance. Recumbent bikes have less wind resistance because of their low profile. These bikes are often equipped with aerodynamic windshields, which further reduce wind resistance. When cyclists ride as a group, they commonly engage in a practice known as drafting, in which they trail each other closely to take advantage of the reduced wind resistance behind other riders. After a turn at the front of the pack, the lead riders drop to the rear, and less fatigued riders take the lead.

## BENEFITS OF CYCLING

Many environmentalists promote the nonpolluting benefits of bicycle use, as well as the fact that bicycles do not consume fossil fuels. Some experts have estimated that if everyone in the United States who lives 16 kilometers (10 miles) or less from work were to travel to work every day on a bicycle rather than in a motor vehicle, the nation could become independent of foreign oil. Cycling is also a pleasant form of exercise that can improve physical fitness, and cycling clubs around the world enhance the enjoyment of this activity for many enthusiasts.

Beyond recreational riding, many cyclists engage in highly competitive activities. Many bicycle races are conducted in many different nations; some take place on paved roads, some on off-road routes, and some on oval tracks called velodromes. One well-known road race is the annual Tour de France, in which professional teams of cyclists race throughout France, including routes that take them up and down the Alps, covering some 3,200 kilometers (2,000 miles) in approximately three weeks.

*Robin L. Wulffson*

## FURTHER READING

Armstrong, Lance. *Comeback 2.0: Up Close and Personal.* New York: Simon & Schuster, 2009.

Hurst, Robert. *The Art of Cycling: A Guide to Bicycling in Twenty-first-Century America.* Guilford, Conn.: Falcon, 2004.

Peveler, Willard. *The Complete Book of Road Cycling and Racing.* Camden, Maine: Ragged Mountain Press, 2008.

Sovndal, Shannon. *Cycling Anatomy.* Champaign, Ill.: Human Kinetics, 2009.

# Biogeochemical cycles

CATEGORIES: Weather and climate; atmosphere and air pollution

DEFINITION: Movements of chemical elements among parts of the earth, including the atmosphere, the earth's crust, oceans, and living things

SIGNIFICANCE: The movement and location of chemical elements among the various systems that make up the earth affect the planet's climate, the diversity and range of species, the impact and intensity of geological events, and many other matters related to the earth as a whole and subsystems within it.

Modern geoscientists consider the earth to comprise a set of interacting open systems: the atmosphere (the layers of air that envelop the earth), the biosphere (the areas that support and are filled with living things), the lithosphere (the earth's crust), and the hydrosphere (the water found on the earth). These various spheres contain within them important chemical elements. The six most abundant elements are carbon, nitrogen, oxygen, hydrogen, phosphorus, and sulfur; these elements, alone or in combination with others, serve as critical macronutrients to living things. Other important macronutrients are potassium, calcium, iron, and magnesium. So-called micronutrients are present in smaller amounts and play critical roles in sustaining life on earth: boron (used by green plants), copper (critical to the functioning of some enzymes), and molybdenum (vital for the functioning of nitrogen-fixing bacteria). Other chemical elements, such as ammonia, also are involved in biogeochemical cycles.

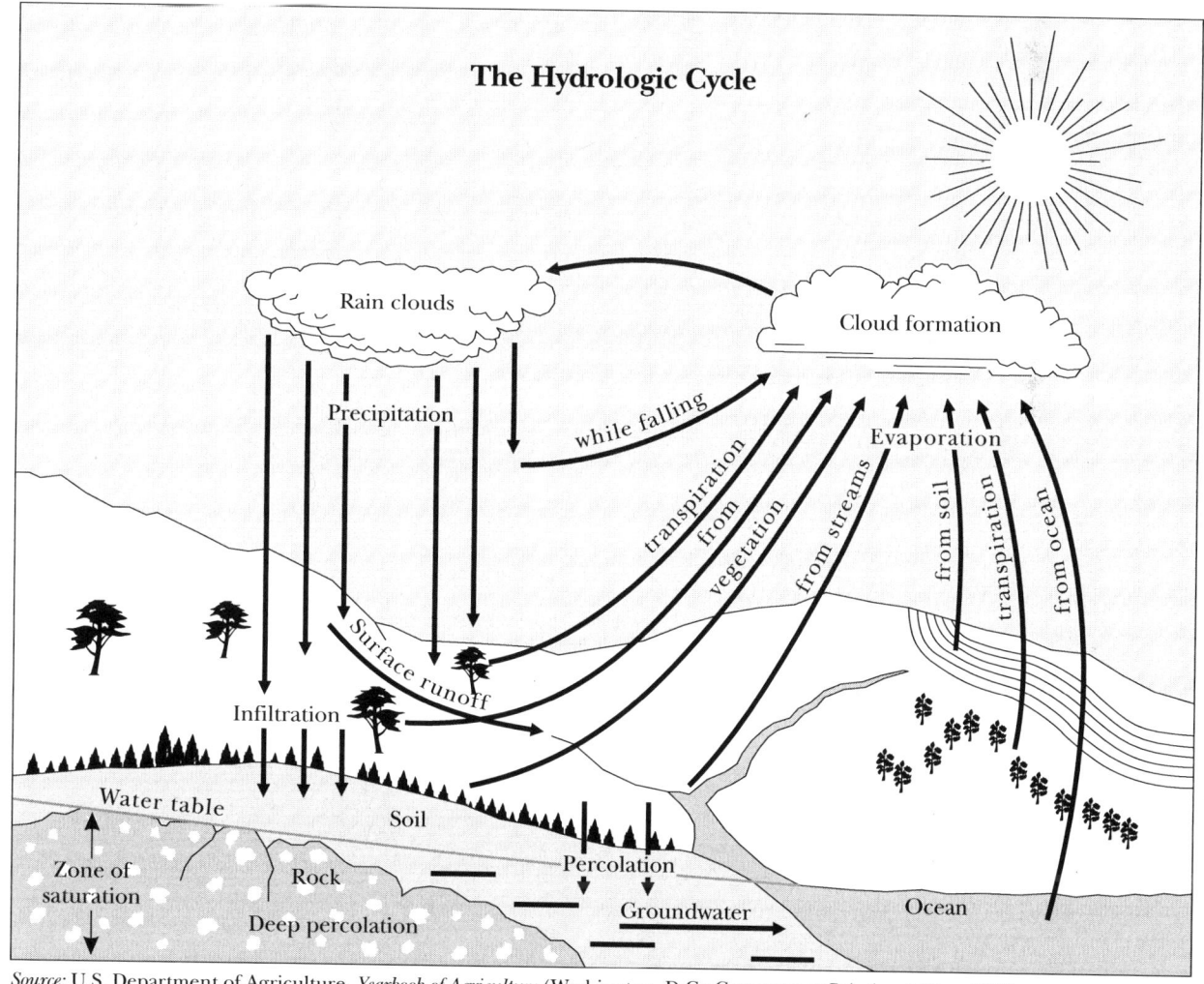

*Source:* U.S. Department of Agriculture, *Yearbook of Agriculture* (Washington, D.C.: Government Printing Office, 1955).

The atmosphere and the oceans serve as important reservoirs where the key gas elements (carbon, nitrogen, oxygen, and hydrogen) are mainly stored. Soils and sedimentary rocks serve as principal reservoirs for the storage of phosphorus and sulfur. Throughout the earth these various chemicals are moved about by different physical, chemical, and human-induced processes that alter the respective amounts of elements in various places and within both living and nonliving things.

An example of a biogeochemical cycle is the hydrologic (water) cycle that moves oxygen and hydrogen, along with other chemicals, from reservoir to reservoir, principally at or near the surface of the earth. This includes the proportionally large amounts of water residing within living things as well as the water found in the oceans, the atmosphere, surface water, groundwater, glaciers, and soils. During various periods of the planet's history the relative percentages of oxygen, hydrogen, and other chemical elements found in the hydrosphere have changed dramatically. These changes have in turn affected the overall temperature and climate of the earth and the types of environments available for living things; they have also sometimes caused physical changes to the other spheres (atmosphere, biosphere, and lithosphere).

Another biogeochemical cycle is the carbon cycle. Most of the original carbon in the earth's systems came out of the earth's mantle in the form of carbon dioxide gas released by volcanoes. The carbon dioxide in the atmosphere is taken up by various means, including dissolution in seawater in the form of bicarbonate ions and absorption by photosynthetic organisms such as algae and plants that convert it into sugar and other organic chemicals. As this carbon enters the food chain it ultimately finds its way into the tissues of animals. It is estimated that some 63 billion tons of carbon move every year worldwide from the atmosphere into life-forms. In this sense the carbon that has been so transformed has moved from the reservoir of the atmosphere into the reservoir of the biosphere.

Human beings' burning of fuels that are rich in carbon has resulted in increases in the amounts of carbon dioxide and other greenhouse gases in the earth's atmosphere, beyond the rate that would have occurred from natural processes. Most climatologists and atmospheric scientists believe that these greenhouse gases contribute to increasing the earth's overall atmospheric temperature and that the resulting net global warming has the potential to produce catastrophic effects over time.

*Dennis W. Cheek*

FURTHER READING

Bashkin, Vladimir N., with Robert W. Howarth. *Modern Biogeochemistry*. Norwell, Mass.: Kluwer Academic, 2003.

Libes, Susan M. *Introduction to Marine Biogeochemistry*. 2d ed. New York: Academic Press, 2009.

# Black lung

CATEGORY: Human health and the environment

DEFINITION: Chronic respiratory disease caused by long-term inhalation of coal and mineral dusts in closed coal-mining environments

SIGNIFICANCE: Improvements of conditions for workers in coal mines have helped to reduce the incidence of coal workers' pneumoconiosis, commonly known as black lung, but the disease has not been eradicated.

Since the Industrial Revolution, coal has provided energy for industrialized societies. With escalating demand, coal mining developed into a prevalent industry in many regions, such as Wales in Great Britain and the Appalachian Mountains of the United States. By the nineteenth century, the economies of such regions were dependent on coal.

The burning of coal produces gaseous and particulate (soot) pollution, and cities in the nineteenth century were characterized by black, soot-coated buildings and coal residue on other surfaces. Explosions caused by coal dust became common in coal mines and storage facilities. Slower to appear from the use of coal for energy were the greenhouse effect, acid rain, and coal workers' pneumoconiosis, also called black lung, caused by chronic inhalation of coal dust. This particle-induced fibrosis-emphysema produces lesions in respiratory bronchioles that interfere with the absorption and transport of oxygen in the lungs. Its symptoms are consistent with similar lung diseases: weakness and poor health, shortness of breath and oxygen starvation, heart disease, immune system irregularities, and lung cancers. It resembles diseases caused by fine airborne, respirable particles of asbestos (asbestosis); cotton, wood, and other plant-based dusts (farmers' lung); and silicas such as quartz, glass, and sand (silicosis).

The greatest incidence of black lung occurs in underground mining. Drilling, pulverizing, loading, and transporting coal generate large dust concentrations, which are breathed into the workers' lungs. Numerous safeguards have been instituted in coal mining to minimize the risks of workers' developing black lung. Mine passages are ventilated by fans and baffles, and sophisticated routing produces multiregion flows that keep the air at face level fresh. Mines also utilize wetting and scrubber systems to keep coal wet and water mists to remove dust from the air.

More recent developments include ventilation and filtering systems worn on the body—for example, helmets that provide continuous streams of filtered air across workers' faces. Research is ongoing in the development of drill bits and other components that can minimize the dust produced. The replacement of steel bits with bits coated with tungsten carbide, polycrystalline diamond, or other hard ceramic or metallic films has received attention, as has optimization of thread geometry and bit speed.

Since the Federal Coal Mine Health and Safety Act of 1969 and the Mine Act amendments of 1977, federal and state agencies (for example, the Mine Safety and Health Administration, or MSHA) have mandated that mine operators undertake efforts aimed at preventing black lung. Operators are required to sample mine air for deviations from permissible exposure limits (PELs) of coal and silica dusts, methane, carbon monoxide, sulfur dioxide, and hydrogen sulfide; to report these deviations; and to take appropriate measures to correct them.

Operators are also required to provide medical screenings for employees over time. Periodic chest X rays, measurements of lung capacity, cardiovascular checkups, and general blood, urine, and endocrine analyses warn of early signs of black lung. Worker health data are gathered through periodic questionnaires, providing insight into the overall risks in a given operation. Emphasis has been placed on education regarding black lung and the regulatory infrastructure that now addresses it. Numerous government agencies offer information and services related to the disease, its prevention, its control, and compensation or support for those affected.

The MSHA's PEL for coal dust is 2 milligrams per cubic meter ($mg/m^3$) for unaffected workers and 1 $mg/m^3$ for workers with any signs of black lung. The National Institute for Occupational Safety and Health (NIOSH) recommends an exposure level no greater than 1 $mg/m^3$ for all workers. The Occupational Safety and Health Administration (OSHA) has set a slightly different PEL of 2.4 $mg/m^3$ for coal dust with less than 5 percent silica. In 1968-1969 the average dust concentration in underground coal mines was 6 $mg/m^3$, but since the Coal Act and the Mine Act, these averages have dropped to below 2 $mg/m^3$.

The silica content of coal dust affects the epidemiology of black lung as well as permissible exposure limits. Less than 1 $mg/m^3$ of silica can cause silicosis, and the current PEL for silicon is 0.1 $mg/m^3$, with reduction to 0.05 $mg/m^3$ being considered. Silica and mineral dusts are generated during initial drilling through rock to reach the coal in coal deposits that are interspersed with bedrock.

Significant progress was made over the last three to four decades of the twentieth century in fighting black lung and associated problems in coal mines. Government intervention, technological developments, and education should further improve the health, safety, and environmental status of the coal mining industry.

*Robert D. Engelken*

### FURTHER READING

Goodell, Jeff. *Big Coal: The Dirty Secret Behind America's Energy Future.* New York: Houghton Mifflin, 2006.

Levine, Linda. *Coal Mine Safety and Health.* Washington, D.C.: Congressional Research Service, 2008.

Meyers, Robert A., ed. *Coal Handbook.* New York: Marcel Dekker, 1981.

Ripley, Earle, Robert Redmann, and Adele Crowder. *Environmental Effects of Mining.* Boca Raton, Fla.: CRC Press, 1996.

Witschi, Hanspeter, and Paul Nettesheim, eds. *Mechanisms in Respiratory Toxicology.* London: Chapman & Hall, 1982.

## Black Wednesday

CATEGORIES: Disasters; atmosphere and air pollution

THE EVENT: Severe episode of photochemical smog in Los Angeles

DATE: September 8, 1943

SIGNIFICANCE: After the Black Wednesday incident helped to demonstrate the extent of the health dangers posed by smog, government actions aimed at addressing the problem were undertaken.

The Los Angeles, California, basin is prototypical of a smog-producing area, with mountains rising on the east and north and persistent high-pressure weather systems during summer and early fall. Subsidence in upper levels of the high-pressure atmosphere results in air being compressed and heated. Westerly winds blowing over a cold ocean current carry cool, moist air in beneath the subsiding air, forming a temperature inversion at an altitude lower than the mountain peaks. With cool, moist air beneath hot, dry air, rising currents can ascend only to the inversion level. They then spread laterally and descend when they run into the natural barrier of the mountains. This situation places a lid over rising air, which causes pollutants to remain at low elevations for several days.

When Los Angeles experienced a severe episode of photochemical smog on September 8, 1943, the *Los Angeles Times* dubbed the day Black Wednesday. The human-made origins of Black Wednesday included increased industrialization and population growth in the Los Angeles area as a result of World War II. These factors led to a rise in the number of automobiles, with an accompanying increase in automotive emissions, as well as a higher yield of effluent from industrial smokestacks. In addition, a shortage of natural rubber led to the manufacture of synthetic rubber. A plant in the Los Angeles neighborhood of Boyle Heights produced a synthetic called butadiene, which appeared to be a significant source of air pollution. Other minor sources contributing to the photochemical smog were backyard incinerators used to burn refuse and the smudge pots, or orchard heaters, used by citrus growers.

The consequences for humans of exposure to photochemical smog can vary. Carbon monoxide (CO) from automotive exhaust can result in a change in physiology: Exposure to low levels of CO can produce impaired functions, whereas exposure to high concentrations may end in death. There seems to be a positive correlation between polyaromatic hydrocarbons and lung, skin, and scrotal cancers. The main complaints of those exposed to smog, however, are burning eyes and irritated throats. Peroxyacetyl nitrates (PAN) are the primary eye irritants, although acrolein and formaldehyde also contribute to the

problem. Finally, reduced visibility is a product of aerosols contained in smog.

Vegetation is distressed by several components of photochemical smog. Ozone assaults leaf palisades, resulting in destruction of chlorophyll, lowered photosynthesis and respiration rates, and development of dark spots on leaf surfaces. Plant exposure to sulfur dioxide causes gas to combine with water to form sulfite ions, which cause leaves to darken, grow flaccid, and become dry. This eventually kills tissue. The effects of photochemical smog on animals were not studied or recorded during the Black Wednesday event, but animals are believed to experience impacts of exposure to smog that are similar to those experienced by humans.

Photochemical smog also alters inanimate objects. Rapid cracking and eventual deterioration of stretched rubber can result from exposure to smog. Likewise, breakdowns of natural and synthetic fabrics and fading of dyes can be traced to smog elements.

In the years following the Black Wednesday incident, government and civic organizations began to take steps aimed at curbing air pollution in Southern California. The use of backyard incinerators was banned, and increasingly strict laws were passed concerning automobile emissions.

*Ralph D. Cross*

FURTHER READING

Carle, David. *Introduction to Air in California*. Berkeley: University of California Press, 2006.

Jacobs, Chip, and William J. Kelly. *Smogtown: The Lung-Burning History of Pollution in Los Angeles*. Woodstock, N.Y.: Overlook Press, 2008.

Vallero, Daniel. *Fundamentals of Air Pollution*. 4th ed. Boston: Elsevier, 2008.

# Carbon cycle

CATEGORIES: Atmosphere and air pollution; weather and climate

DEFINITION: Pathways by which carbon moves through the environment

SIGNIFICANCE: The balance of the carbon cycle determines the atmospheric concentration of carbon dioxide, which, through its role as a greenhouse gas, modulates the earth's temperature. Human activities that affect the carbon cycle thus have an effect on global climate.

Carbon is naturally exchanged between the atmosphere and the oceans, the terrestrial biosphere and soils, and the solid earth. The preindustrial balance of these carbon exchanges led to an atmospheric carbon dioxide concentration of 280 parts per million. Human industrial activities have added carbon dioxide to the atmosphere, increasing the atmospheric concentration to 380 parts per million.

Carbon is naturally removed from the atmosphere by photosynthesis on land and by dissolution of atmospheric carbon dioxide in the oceans. Carbon is introduced into the atmosphere by respiration and combustion of terrestrial organic matter, by outgassing of carbon dioxide from the oceans, and as a by-product of human industrial activities. Terrestrial plants remove about 15 percent of the atmosphere's carbon dioxide each year through photosynthesis. About half of this carbon is respired by these same plants as they release energy for their internal metabolic processes. The other half of the carbon removed from the atmosphere by terrestrial photosynthesis is primarily re-

turned to the atmosphere through respiration of organic matter by decomposers. Most of this carbon is returned to the atmosphere as carbon dioxide through aerobic respiration, but in low-oxygen wetland environments carbon can be respired anaerobically and released to the atmosphere as methane, a much more potent greenhouse gas. This methane is then broken down to carbon dioxide in the atmosphere over a timescale of about eight years.

Approximately 12 percent of the atmosphere's carbon dioxide is exchanged with the oceans each year. Carbon is more soluble in cold water than in hot water, and so the oceans take up carbon at high northern latitudes and emit carbon in the Tropics. Photosynthesis in the surface ocean binds dissolved carbon into organic matter. Much of this carbon is returned to the surface ocean by breakdown of this organic matter, but some of this material is packaged into large enough clumps of organic matter that it sinks under its own weight and is redissolved in the deep ocean. This process, known as the biological pump,

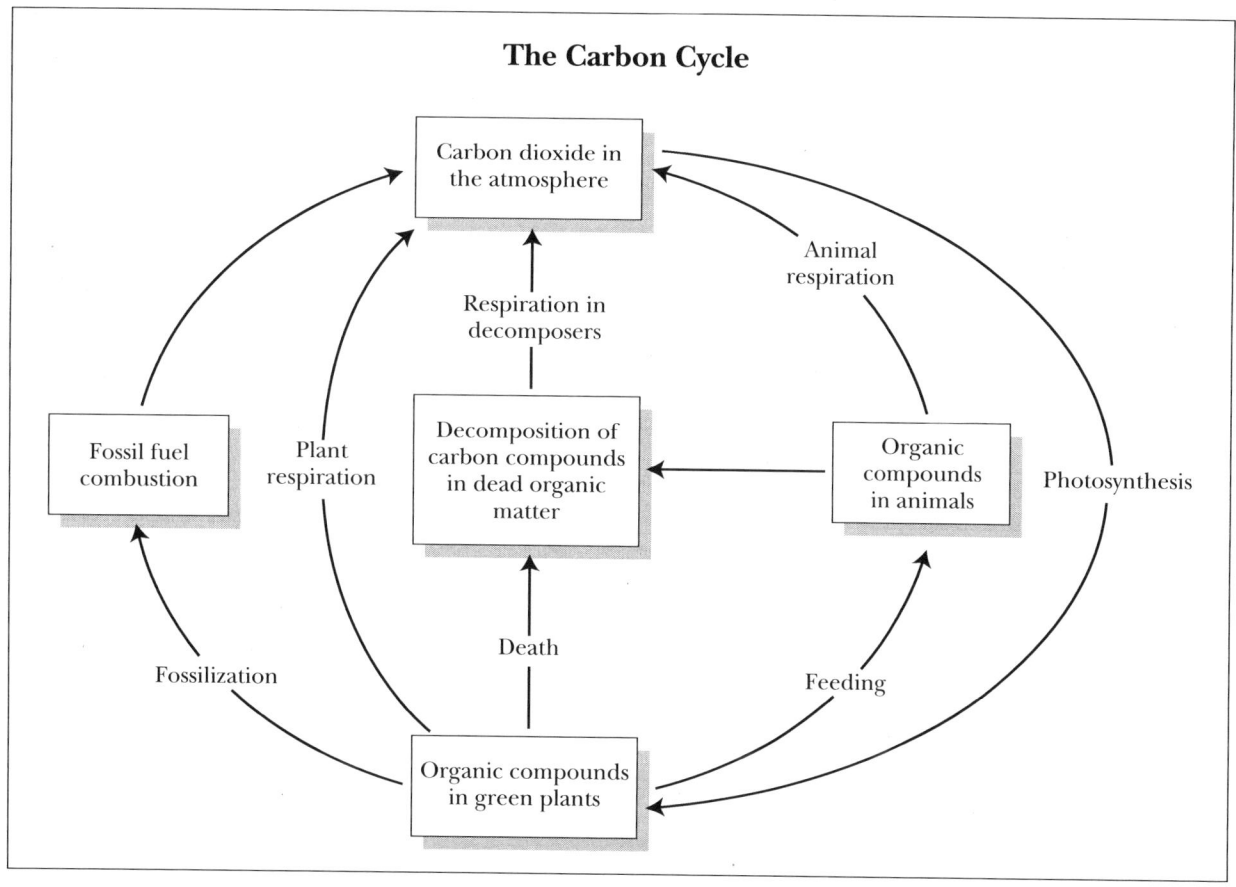

acts to move carbon (and nutrients) from the surface ocean to depth. In ocean upwelling zones, carbon-rich water from depth is brought up to the surface, producing a source of carbon to the atmosphere. Exchanges of carbon between the solid earth and the other reservoirs are important only on geologic timescales.

Human industrial activity introduces an amount of carbon to the atmosphere each year equivalent to about 1 percent of the total atmospheric carbon content. The rate of atmospheric carbon dioxide increase is, however, only half the rate at which industrial activity introduces carbon dioxide to the atmosphere. The other half is removed from the atmosphere by dissolution in the oceans and uptake by the land surfaces.

Carbon-cycle models predict that global warming will trigger processes that will alter the natural carbon cycle in a way that will decrease the ability of natural processes to remove industrial carbon from the atmosphere and accelerate the rate of increase of atmospheric greenhouse gases. Warming of the oceans decreases the solubility of carbon dioxide in water. Decomposition of organic material in soils to produce carbon dioxide proceeds more rapidly at higher temperatures. Permanently frozen soils at high latitudes (permafrost) globally hold more carbon than the atmosphere, and this carbon becomes exposed to decomposition when the permafrost thaws.

*Alexander R. Stine*

FURTHER READING

Field, Christopher B., and Michael R. Raupach, eds. *The Global Carbon Cycle: Integrating Humans, Climate, and the Natural World.* Washington, D.C.: Island Press, 2004.
Houghton, R. A. "The Contemporary Carbon Cycle." *Treatise on Geochemistry* 8, no. 10 (2003): 473-513.
Wigley, T. M. L., and D. S. Schimel, eds. *The Carbon Cycle.* New York: Cambridge University Press, 2000.

# Carbon dioxide

CATEGORIES: Pollutants and toxins; atmosphere and air pollution

DEFINITION: Chemical compound in which molecules are composed of one carbon atom and two oxygen atoms

SIGNIFICANCE: The increase of carbon dioxide in the earth's atmosphere that has been occurring since the Industrial Revolution, if not earlier, has been linked to global warming. International efforts undertaken to limit emissions of carbon dioxide have met with varying degrees of success.

Carbon dioxide ($CO_2$), a gas that is essential to life on earth, is generated by the burning of fossil fuels such as wood, oil, and coal and by the decomposition of organic matter. Approximately 57 percent of greenhouse gases, of which $CO_2$ is one, come from the burning of fossil fuels. Land clearing also contributes to $CO_2$ in the atmosphere. A large amount of $CO_2$ is found in deposits in the floors of the earth's oceans. When this gas is released into the atmosphere, it remains and helps to trap solar radiation in a process that has been shown to contribute to climate change.

The amount of carbon dioxide in the atmosphere has varied over the life of the planet; the amount has been both much lower and much higher than that present during the early twenty-first century. The $CO_2$ level in the atmosphere was relatively stable, however, from the end of the last ice age until the nineteenth century, averaging 280 parts per million (ppm). From the start of the Industrial Revolution to the middle of the twentieth century, $CO_2$ levels grew at a steady rate, and the rate of growth increased in the latter half of the twentieth century. When measurements were first undertaken at the Mauna Loa Observatory in Hawaii in the 1950's, the measured concentration of $CO_2$ was 315 ppm; by 2008 the concentration stood at 385 ppm. Some scientists estimate that the amount of $CO_2$ in the earth's atmosphere could be as high as 1,000 ppm by the year 2100 if emissions continue to increase unchecked. Such a level could produce global temperatures that are 5 degrees Celsius (9 degrees Fahrenheit) warmer than in the early twenty-first century, a level not seen for several million years. Even a moderate increase to 440 ppm is expected to produce a temperature increase of 3 degrees Celsius (5.4 degrees Fahrenheit).

Industrialized countries such as the United States and Germany are major producers of all greenhouse gases, especially $CO_2$, owing to their high levels of use of fossil fuels. Some industrialized nations have turned increasingly to energy sources that produce less $CO_2$, such as natural gas. However, industrializing nations such as India and China are also major producers of $CO_2$, largely because they burn a great deal of coal, which produces more $CO_2$ than other fossil fuels.

Somewhat surprisingly, the industrializing nation of Brazil is a significant producer of $CO_2$ as the result of the clearance of massive areas of land in the Amazon basin for agriculture; this process produces $CO_2$ as the biomass decomposes or is burned. Other, less industrialized nations in Asia, Africa, and South America also produce $CO_2$ as they harvest large amounts of timber and clear land for agriculture or mining.

The Kyoto Protocol of 1997, an international agreement aimed at limiting and reducing $CO_2$ emissions, applied only to industrialized countries and countries such as China and India that have exhibited the fastest rate of growth in $CO_2$ emissions. Although some countries have tried to reduce $CO_2$ emissions, the United States initially refused to ratify the agreement and, during the presidential administration of George W. Bush, was slow to try to limit emissions, arguing that industrial production was more important than potential climate change. The delegates to the 2009 United Nations Climate Change Conference, held in Copenhagen, Denmark, attempted to address the issue of regulating $CO_2$ and other greenhouse gas emissions further, but little was accomplished.

*John M. Theilmann*

FURTHER READING

Houghton, John. *Global Warming*. 3d ed. New York: Cambridge University Press, 2004.
Smail, Vaclav. *Cycles of Life*. New York: Scientific American, 1997.
Volk, James. *CO₂ Rising*. Cambridge, Mass.: MIT Press, 2008.

# Carbon dioxide air capture

CATEGORY: Atmosphere and air pollution
DEFINITION: The trapping or elimination of carbon dioxide emitted from industrial or commercial sources before it can enter the atmosphere
SIGNIFICANCE: Carbon dioxide capture and sequestration or storage is anticipated to play a bridging role between carbon dependence and a sustainable low-carbon energy future by serving as the critical enabling technology that will lead to a signifi-

cant reduction in $CO_2$ release into the air while allowing industrial processes such as the burning of coal for power generation to continue to meet global energy needs.

The term "carbon dioxide air capture" is used to describe a set of technologies aimed at preventing the carbon dioxide ($CO_2$) emitted from industrial or commercial sources from entering the atmosphere. The processes involved are also commonly referred to as carbon dioxide capture and sequestration or carbon capture and storage (CCS). The geoengineering technique of scrubbing (absorbing) $CO_2$ from ambient air is also sometimes referred to as CSS, as are biological techniques that employ organisms (such as plankton) and organic matter to capture $CO_2$ from the air.

In CCS, the process of capturing $CO_2$ from a large emissions source is often coupled with the subsequent compression of the gas, storage of the gas (for example, by injecting it into deep underground geological formations or into deep ocean masses called saline aquifers, or by converting it into the form of mineral carbonate), and recycling of the gas to enhance industrial processes, such as that seen in $CO_2$-assisted enhanced oil recovery (EOR), in which $CO_2$ gas is injected into an oil-bearing stratum under high pressure to cause oil to be displaced upward.

The CCS process begins with the capture of $CO_2$ generated by a power station or other industrial facility, such as a cement factory, steelworks, or oil refinery. The $CO_2$ can be captured before, during, or after the source's combustion (burning) of fossil fuels. Precombustion capture involves the separation of fossil fuels into hydrogen and $CO_2$ before they are burned. For example, in the instance of coal, the process involves the conversion of coal into a synthetic gas (syngas) made up of carbon monoxide and hydrogen. The syngas is reacted further with steam to produce a $CO_2$-hydrogen mix. Further processing produces a mix with a high concentration of $CO_2$, which is separated out. The remaining hydrogen is then utilized as a $CO_2$-free energy source that produces only heat and water vapor when combusted. Precombustion capture technology has been widely implemented in the fertilizer industry as well as in natural gas forming.

In oxy-fuel combustion, burning of the fossil fuel in oxygen instead of air results in an exhaust gas that is $CO_2$-free. This technology is commonly used in the glass furnace industry. In postcombustion capture, the $CO_2$ is separated from flue (exhaust) gases after the combustion of fossil fuels. The $CO_2$ content is usually much lower than in the gas that is separated during oxy-fuel combustion or precombustion capture, with the volume of $CO_2$ in the range of 3 percent to 15 percent by volume.

## CHALLENGES TO IMPLEMENTATION

One challenge facing CCS is the demonstration of its efficacy and safety on an industrial scale at competitive cost. While CCS is known to be safe and is well understood in terms of the fundamental science and technical requirements, no evidence has been gathered regarding the process's long-term impacts on the environment (for example, the safety of storing $CO_2$ in geological formations) or possible danger to humans (for instance, if $CO_2$ leaks from storage).

CCS applied to a modern power plant could reduce $CO_2$ emissions by up to 90 percent compared to an equivalent plant with no CCS devices, but the implementation of $CO_2$ capture is significantly more expensive than the use of traditional systems of emissions control. For example, capturing and compressing $CO_2$ increases the fuel requirement of a coal-fired plant by as much as 25-40 percent. It is estimated that the cost of energy produced by a new power plant with CCS is from 21 percent to 91 percent higher than that produced by a non-CCS power plant.

Aside from the cost and technical challenges of CCS, a regulatory framework needs to be established to support CCS and to clarify at regional, national, and international levels the long-term rights, liabilities, and technical requirements associated with the use of CCS technologies. Moreover, before investors, scientists, politicians, and industries can be persuaded that CCS is a worthwhile investment, agreements need to be reached on a price on carbon emissions and on whether a carbon tax, cap-and-trade regime, or other carbon-trading/taxation framework will be implemented.

*Rena Christina Tabata*

## FURTHER READING

Hanjalić, K., R. van de Krol, and A. Lekić, eds. *Sustainable Energy Technologies: Options and Prospects*. London: Springer, 2008.

Kutz, Myer. *Environmentally Conscious Fossil Energy Production*. Hoboken, N.J.: John Wiley & Sons, 2009.

Rackley, Steve. *Carbon Capture and Storage*. Boston: Butterworth-Heinemann, 2009.

Rojey, Alexandre. *Energy and Climate: How to Achieve a Successful Energy Transition*. Chichester, England: John Wiley & Sons, 2009.

Shiosani, Fereidoon P. *Generating Electricity in a Carbon-Constrained World*. Burlington, Mass.: Academic Press, 2009.

# Carbon monoxide

CATEGORIES: Pollutants and toxins; atmosphere and air pollution

DEFINITION: Odorless and colorless toxic gas in which molecules consist of one carbon atom and one oxygen atom

SIGNIFICANCE: The air pollutant carbon monoxide is a component of automobile exhaust emissions, but the gas is perhaps most dangerous indoors, where it can cause accidental death by asphyxiation.

Carbon monoxide (CO), an indoor and outdoor air pollutant, is formed by incomplete combustion of fossil fuels when insufficient oxygen is present to convert carbon compounds to nontoxic carbon dioxide. Indoors, CO may come from unvented kerosene and gas space heaters; leaking or improperly vented chimneys, furnaces, gas water heaters, wood stoves, gas stoves, and fireplaces; generators and other gasoline-powered equipment; automobile exhaust from attached garages; and tobacco smoke. Outdoors, CO is present in small amounts naturally in the atmosphere, chiefly as a product of volcanic activity, but also from naturally occurring fires.

Motor vehicle exhaust contributes about 56 percent of all CO emissions in the United States. The highest levels of CO occur in areas with heavy traffic congestion, and in cities, 85-95 percent of all CO emissions may come from motor vehicle exhaust. Other nonroad engines and vehicles (such as construction equipment and boats) contribute about 22 percent of CO emissions nationwide. Additional sources of CO emissions include industrial processes (metals processing and chemical manufacturing) and residential wood burning. The highest levels of outdoor CO typically occur during the winter months, for three primary reasons: Motor vehicles need more fuel to start at cold temperatures, some emissions-control devices (oxygen sensors and catalytic converters) operate less efficiently in the cold, and temperature inversion conditions, which trap CO near the ground beneath a layer of warm air, are more frequent.

CO enters the bloodstream through the lungs and binds to the blood protein hemoglobin, the main function of which is to transport oxygen to body tissues, including vital organs such as the heart and brain. The CO and hemoglobin combine to form the compound carboxyhemoglobin, and the hemoglobin is no longer available for transporting oxygen. In closed indoor environments, the level of carboxyhemoglobin in the body can rise to toxic levels and eventually result in asphyxiation.

People with heart disease are especially sensitive to CO poisoning because of its reduction of oxygen transport through the blood; they may experience chest pain if they breathe the gas. Infants, elderly persons, and individuals with respiratory diseases are also particularly sensitive to CO. CO can affect healthy individuals by impairing exercise capacity, visual perception, manual dexterity, learning ability, and the ability to perform complex tasks.

During the early 1970's, the U.S. Environmental Protection Agency (EPA) set two national air-quality standards for CO in motor vehicle emissions: a one-hour standard of 35 parts per million (the highest allowable level of CO measured in ambient air over a one-hour period) and an eight-hour standard of 9 parts per million. The EPA also mandated reductions in emissions from large industrial facilities. Advances in vehicle technologies and the development of cleaner fuels have resulted in reductions in CO emissions. Across the United States, air-quality stations measure the levels of CO (and other pollutants) in the air on a regular basis. The measurements are compared with the EPA standards, and areas that have CO levels that exceed the standards are required to develop and carry out plans to reduce their CO emissions. As a result of these steps, CO emissions from automobiles, motorcycles, and light- and heavy-duty trucks have been reduced by more than 40 percent since 1970. The greatest reduction has been the nearly 60 percent decrease in automobile emissions of CO.

*Bernard Jacobson*

FURTHER READING

Kleinman, Michael T. "Carbon Monoxide." In *Environmental Toxicants: Human Exposures and Their Health Effects*, edited by Morton Lippmann. 3d ed. Hoboken, N.J.: John Wiley & Sons, 2009.
Penney, David G., ed. *Carbon Monoxide Poisoning*. Boca Raton, Fla.: CRC Press, 2008.

# Carcinogens

CATEGORIES: Human health and the environment; pollutants and toxins
DEFINITION: Substances or physical agents that cause or worsen cancer
SIGNIFICANCE: The effects of human exposure to carcinogens in the environment may include the development of different types of illness, deaths, and economic obligations on a national and global scale.

Cancer is a leading cause of death throughout the world. Environmentalists and others have raised concerns regarding the cancer-causing (carcinogenic) potential of exposure to a constantly growing number of both newly developed and long-existing chemicals in the environment. In addition, humans seem to be increasingly exposed to various sources of electromagnetic waves, such as microwaves, and some groups and individuals are concerned about the possible carcinogenicity of these physical phenomena.

Carcinogens can be categorized based on their origin as chemicals (naturally occurring or synthetic), physical agents, or infectious agents. Chemical carcinogens can be classed as compounds that occur naturally, such as aflatoxins, chromium 6 compounds, and arsenic compounds; and others that are largely synthetic in origin, such as benzene, formaldehyde, vinyl chloride, and dioxins. In addition there are carcinogenic minerals, such as asbestos. Other carcinogenic chemicals are elements or substances such as radon (a radioactive gas), beryllium, and cadmium. Some carcinogens—such as tobacco smoke and alcoholic beverages—are mixtures of compounds. Physical agents that are carcinogens include solar radiation (primarily ultraviolet radiation), gamma rays, and X rays.

Infectious disease agents that have been implicated as carcinogens include human papilloma virus (HPV), which can cause cervical cancer; *Helicobacter pylori* (*H. pylori*), a bacterium causally associated with stomach cancer; the hepatitis C and hepatitis B viruses, which can cause liver cancer; Epstein-Barr virus, which is associated with Burkitt's lymphoma; and human T-lymphotrophic virus type 1 (HTLV-1), which has been linked to leukemia in adults. The human immunodeficiency virus (HIV), which causes acquired immunodeficiency syndrome (AIDS), has been associated with Kaposi's sarcoma. Viruses have also been shown definitively to cause tumors in animals such as mice (mammary tumor virus), chickens (Rous sarcoma virus), and Tasmanian devils.

Exposure to carcinogens can be related to work environments, such as in the case of workers in the nuclear power and medical radioisotope industries. Sometimes carcinogenic agents happen to be concentrated in particular geographic regions; for example, widespread exposure to the carcinogen arsenic occurred in Bangladesh as the result of contaminated drinking water from a large number of wells that accessed groundwater in which arsenic was uncommonly abundant. Exposure to *H. pylori* is believed to occur through contaminated water supplies and thus is considered to be environmental in origin. Tobacco smoke, a major cause of lung cancer, is an environmental carcinogen to which people are widely exposed.

Carcinogens may lead to cancer by directly damaging deoxyribonucleic acid (DNA), as in the case of radiation; through conversion through metabolism; or through effects on metabolism. In the case of viruses, viral genetic material may be incorporated into host DNA at sites of oncogenes, which are genes whose altered function or disruption leads to cancer. Different carcinogens often lead to effects on different organs, thus *H. pylori* is associated mainly with stomach cancer, whereas tobacco smoking or use is associated with lung, oral, and laryngeal cancers (and also bladder, colon, and kidney cancers, among others), and asbestos is primarily linked to lung cancer. The time period and frequency of exposure to carcinogens as well as a person's genetic background can also influence the likelihood that a carcinogen causes cancer.

*Oluseyi A. Vanderpuye*

FURTHER READING

Hill, Marquita K. "Chemical Exposures and Risk Assessment." In *Understanding Environmental Pollution*. 3d ed. New York: Cambridge University Press, 2010.

McKinnell, Robert. G., et al. *The Biological Basis of Cancer*. 2d ed. New York: Cambridge University Press, 2006.

Ward, Elizabeth M. "Cancer." In *Occupational and Environmental Health: Recognizing and Preventing Disease and Injury*, edited by Barry S. Levy et al. 5th ed. Philadelphia: Lippincott Williams & Wilkins, 2006.

## Cash for Clunkers

CATEGORY: Atmosphere and air pollution

IDENTIFICATION: U.S. federal government program that briefly subsidized purchases of cleaner-running and more fuel-efficient vehicles

DATES: June 24-August 24, 2009

SIGNIFICANCE: The three goals of the Cash for Clunkers program were to help reduce American dependence on foreign oil, help reduce carbon emissions from passenger vehicles, and provide a stimulus to the American automobile industry. It at least partly achieved the first two goals, but claims about its success in achieving the third goal have been challenged.

Enacted on June 24, 2009, and officially known as the Car Allowance Rebate System (CARS), the popularly called Cash for Clunkers program made a total of $3 billion available to subsidize the private purchase of more fuel-efficient automobiles, trucks, sport utility vehicles (SUVs), and vans before it was terminated ahead of schedule on August 24 because its funding had been exhausted by popular demand. The subsidies were disbursed as credits toward the purchase of only new vehicles and, depending on the magnitude of the improvement in miles per gallon (mpg) offered by the new vehicle, the individual credits were either $4,500 (for major mpg gains) or $3,500 (for lesser gains). New purchases had to cost less than $45,000, trade-ins generally had to have combined city and highway mileage ratings of 18 mpg or less, and all trade-ins were to be scrapped, not resold.

A similar program had been pursued previously by the Slovak Republic in Europe, and other countries were considering adopting such progams when the U.S. Congress enacted Cash for Clunkers. The program's goals were to stimulate sales in the sagging American auto industry, to reduce American dependence on foreign energy by replacing "gas guzzlers" with more fuel-efficient vehicles, and to reduce carbon emissions emanating from older, fossil-fuel-burning vehicle engines.

The program unquestionably had some success in achieving the second and third of these goals. By definition, replacing low-mileage vehicles with higher-mileage vehicles, all else (such as driving distances) remaining even, reduced the U.S. demand for gasoline and resulted in lower carbon emissions per car on the road. To the extent that some of those cars were also hybrids, which run part of the time on their self-charging batteries, the benefit was enhanced. However, critics correctly noted that the program only subsidized at most slightly more than 690,000 dealer transactions involving more fuel-efficient vehicles—a proverbial drop in the ocean for a country with more than 250 million registered passenger vehicles. Morever, many American drivers were already being motivated to acquire more fuel-efficient vehicles by steep rises in gasoline prices in the months preceding initiation of the program.

*Joseph R. Rudolph, Jr.*

## Catalytic converters

CATEGORY: Atmosphere and air pollution
DEFINITION: Devices that convert carbon monoxide, hydrocarbons, and nitrogen oxides in automobile exhaust gases into less harmful substances
SIGNIFICANCE: The use of catalytic converters has significantly reduced air pollution caused by automobile exhaust.

Since their invention in the late nineteenth century, automobiles have revolutionized society. However, automobiles represented a significant new source of air pollution, particularly in urban areas. As early as 1943, the effects of automobile exhaust emissions on air quality were noted in the Los Angeles area. In 1952 A. J. Haagen-Smit showed that the interaction of nitrogen oxides and hydrocarbons from automobile exhaust with sunlight results in the formation of secondary pollutants, such as peroxyacetyl nitrates (PAN) and ozone. This new form of air pollution, termed photochemical or Los Angeles smog, was soon recognized as a major pollution problem in urban atmospheres.

In 1961 the state of California began regulating the release of pollutants from automobiles. Two years later, the first federal emission standards were imposed as part of the 1963 Clean Air Act. As further restrictions on automobile emissions were introduced in the 1960's and early 1970's, new devices were developed to aid in reducing emissions. Of these, the catalytic converter was the most important.

The first type of catalytic converter consisted of an inert ceramic support material coated with a thin layer of platinum or palladium metal. Carbon monoxide and hydrocarbons from the exhaust gases attached themselves to the metal surface, where they reacted with molecular oxygen and were converted into carbon dioxide and water. The use of a metal catalyst lowered the temperature and increased the rate at which reaction occurred, making possible the removal of almost all of these two pollutants from the exhaust gases.

Because lead, phosphorus, and other substances can coat the metal surfaces in catalytic converters and prevent them from functioning, the Environmental Protection Agency (EPA) was granted the authority to regulate gasoline composition and fuel additives as part of the 1970 Clean Air Act amendments. A consequence of the introduction of catalytic converters in new automobiles in the mid-1970's was the gradual switch from leaded to unleaded gasoline. As a result, emission of lead from transportation, the major source of lead as an air pollutant, decreased by 97 percent between 1978 and 1987.

While the first-generation catalytic converters dramatically reduced emission of carbon monoxide and hydrocarbons from automobiles, they had no effect on nitrogen oxide emissions. Dual-bed catalytic converters, developed in the early 1980's, use a two-step process to reduce emission of nitrogen oxides, carbon monoxide, and hydrocarbons. Dual-bed catalytic converters were followed by three-way converters, which use platinum and rhodium as the metal catalyst. A feedback system with an oxygen sensor is used to adjust the mixture of air and gasoline sent to the automobile engine to ensure maximum removal of pollutants.

The introduction of catalytic converters significantly reduced the release of pollutants by automobiles. In the United States, emission of carbon monoxide and hydrocarbons from transportation sources decreased by more than 50 percent between 1970 and 1995, even though the number of cars and trucks more than doubled. Similar reductions in automobile pollution have occurred in other countries where catalytic converters have been introduced.

*Jeffrey A. Joens*

FURTHER READING

Halderman, James D., and Jim Linder. *Automotive Fuel and Emissions Control Systems.* Upper Saddle River, N.J.: Pearson/Prentice Hall, 2006.

Yamagata, Hiroshi. *The Science and Technology of Materials in Automotive Engines.* Boca Raton, Fla.: CRC Press, 2005.

# Center for Health, Environment, and Justice

CATEGORIES: Organizations and agencies; human health and the environment

IDENTIFICATION: American nonprofit organization established to assist local communities with environmental issues

DATE: Founded in 1981

SIGNIFICANCE: When it was founded in 1981 as the Citizens Clearinghouse for Hazardous Waste, the organization that became the Center for Health, Environment, and Justice was part of an emerging grassroots movement among people concerned about protecting their local communities from the harmful consequences of environmental hazards.

During the late 1970's Lois Gibbs discovered that her child's school in the Love Canal neighborhood of Niagara Falls, New York, was built on thousands of tons of toxic chemicals. At the time no national organization existed to help communities with environmental issues. With no one to turn to for help, Gibbs organized the Love Canal Homeowners Association in 1978 to protest the situation in her neighborhood. This experience and the legal battles that followed led Gibbs to found the Citizens Clearinghouse for Hazardous Waste in 1981; the organization would later change its name to the Center for Health, Environment, and Justice (CHEJ). CHEJ subsequently grew into a national organization with Gibbs serving as executive director. Its main office is located in Falls Church, Virginia.

Since its establishment, CHEJ has remained a grassroots organization that focuses on helping communities coordinate their responses to environmental hazards. CHEJ seeks to assist local neighborhoods by providing necessary aid, information, resources, training, and strategic or technical assistance. In this way, the organization encourages individuals and communities to take social and political action.

CHEJ's stated mission is to "build healthy communities, with social justice, economic well-being, and democratic governance." This includes protecting consumers from hazardous or toxic products. CHEJ addresses its objectives through various campaigns and programs. Its BE SAFE campaign is a precautionary effort to prevent pollution, and its Child Proofing Our Communities campaign helps to educate communities about strategies for protecting children from environmental hazards. CHEJ's Green Flag Schools Program for Environmental Leadership works with schools to educate students about how they can engage in environmental advocacy. PVC: The Poison Plastic is an example of a CHEJ campaign for safe and healthy consumer products. This ongoing national campaign is aimed at moving major corporations away from using PVC plastic, a substance that is harmful for both the environment and human health.

*Jeff Cervantez*

# Chicago Climate Exchange

CATEGORIES: Organizations and agencies; weather and climate

IDENTIFICATION: Financial institution that operates an emissions allowance trading system in Chicago, Illinois

DATE: Established in 2000

SIGNIFICANCE: The Chicago Climate Exchange operates the only cap-and-trade system covering all six greenhouse gases (carbon dioxide, methane, nitrous oxide, sulfur hexafluoride, perfluorocarbons, and hydrofluorocarbons) and offset projects in North America and Brazil.

The Chicago Climate Exchange (CCX) was founded in 2000 and began operating its emissions allowance trading system, or cap-and-trade system, in 2003; by 2010 CCX had four hundred members. CCX membership comprises organizations that produce greenhouse gas emissions (both big emitters and negligible emitters such as office-based businesses and small institutions) as well as owners of title to qualifying emissions-offset projects that sequester, destroy, or reduce such emissions. Each registered emitter makes a voluntary but legally binding commitment to an emissions reduction schedule that includes a commitment to reduce aggregate emissions by a certain percentage below a set baseline level.

Emission baselines, annual reduction commitments, and offset projects are subject to annual audit by third-party experts. In order to reach their emissions targets, members can reduce emissions through their operational practices (such as by changing the fuels they use, making efficiency improvements, or instituting managerial changes), purchase additional emission allowances from other members who have reduced their own emissions by more than the annual reduction requirement, or purchase offset credits from registered emission reduction projects. Entities and individuals can also trade emissions allowances for the purposes of financial investment.

Exchanges affiliated with CCX include the European Climate Exchange (ECX), an exchange operator in the European Union Emissions Trading Scheme); the Insurance Futures Exchange (IFEX), an exchange platform that trades insurance-based derivatives; the Montreal Climate Exchange (MCeX), a joint venture with the Montreal Bourse to host Canadian emissions allowance trading; Tianjin Climate Exchange (TCX), a joint venture with the China National Petroleum Assets Management Company and the Tianjin Property Rights Exchange facilitating emissions trading in China; and Envex, a joint venture of Climate Exchange PLC and Macquarie Capital Group specializing in environmental markets in Australia and the Asia Pacific region.

*Rena Christina Tabata*

# Chlorofluorocarbons

CATEGORY: Atmosphere and air pollution
DEFINITION: Family of chemical compounds used in air conditioners, refrigerators, and aerosol spray cans
SIGNIFICANCE: Concerns about the destruction of stratospheric ozone by chlorofluorocarbons led to a worldwide ban on the manufacture and use of these compounds.

Chlorofluorocarbons (CFCs) are organic molecules containing chlorine, fluorine, and carbon atoms. The first CFCs were discovered by Thomas Midgley, Jr., in 1928. Because these molecules are chemically inert and easily liquefied, CFCs soon became the standard coolants in refrigerators and air conditioners. They also became widely used as propellants in aerosol spray cans. By 1968, 2.3 billion aerosol cans containing CFCs had been sold in the United States.

In 1970 the British scientist James Lovelock determined that most CFCs entering the atmosphere remained there without significant decomposition. Three years later, Frank Sherwood Rowland and Mario Molina, working at the University of California at Irvine, suggested that CFCs would eventually migrate into the stratosphere. Once there, absorption of ultraviolet light would cause CFCs to release chlorine atoms, which would then react catalytically to remove ozone. Since ozone in the stratosphere prevents high-energy ultraviolet light from reaching the surface of the earth, any decrease in ozone would lead to increased exposure to ultraviolet light on the earth's surface, causing higher levels of skin cancer in humans and damage to plants and animals.

Although evidence from laboratory studies suggested that CFCs in the atmosphere would cause depletion of stratospheric ozone, uncertainty remained as to the degree of ozone destruction that would occur. Nevertheless, in 1975 Oregon became the first U.S. state to ban CFCs in aerosol spray cans. Several other states took similar actions, and in 1977 the Food and Drug Administration (FDA) implemented a ban on the use of CFCs as aerosol propellants to be phased in over a two-year period. Continued uncertainties in predictions of ozone loss and the lack of direct evidence for ozone depletion kept most other countries from restricting the use of CFCs. While the U.S. Environmental Protection Agency (EPA) discussed instituting a total ban on CFCs, no action was taken, in part because of the difficulty in finding adequate substitutes for CFCs.

In 1985 a team of British scientists led by Joseph Farman announced the discovery of significant loss of ozone over the Antarctic. Beginning in the early 1970's, springtime levels of ozone had slowly decreased. By 1985 as much as 40 percent of the ozone usually present in the Antarctic stratosphere during the spring had disappeared. In addition, both the duration and the geographic extent of this ozone hole were increasing. Evidence linking formation of the ozone hole to CFCs in the atmosphere was quickly found.

The discovery of the ozone hole led to further restrictions on CFCs. In 1987 an international agreement called the Montreal Protocol was reached to ban the manufacture and use of CFCs by the year 2010. In the United States, passage of the 1990 Clean Air Act amendments resulted in an accelerated timetable for restrictions on CFCs and related compounds. By the mid-1990's levels of CFCs in the atmosphere had stabilized, and CFCs are expected to disappear gradually from the atmosphere over the next century.

*Jeffrey A. Joens*

FURTHER READING

Joesten, Melvin D., John L. Hogg, and Mary E. Castellion. "Chlorofluorocarbons and the Ozone Layer." In *The World of Chemistry: Essentials.* 4th ed. Belmont, Calif.: Thomson Brooks/Cole, 2007.

Newman, Michael C., and Michael A. Unger. "Environmental Contaminants." In *Fundamentals of Ecotoxicology.* 2d ed. Boca Raton, Fla.: CRC Press, 2003.

Parson, Edward A. *Protecting the Ozone Layer: Science and Strategy.* New York: Oxford University Press, 2003.

# Clean Air Act and amendments

CATEGORIES: Treaties, laws, and court cases; atmosphere and air pollution

THE LAWS: U.S. federal laws that govern standards for air quality

DATES: Enacted on December 17, 1963; amended 1970, 1977, and 1990

SIGNIFICANCE: The Clean Air Act of 1963 and its amendments federalize the regulation of air pollution in the United States to a large degree. The act provides guidelines for minimum standards of air quality as well as maximum levels for the emissions of pollutants. It has served as a model for other federal environmental legislation.

Since the 1880's state and local governments in the United States have put limits on smoke emissions and other forms of air pollution. Federal regulation of the problem, however, did not really begin until 1955, when the Air Pollution Control Act authorized the federal government to conduct research and provide assistance to state and local governments. This act included no national standards, and it ceded responsibility for controlling air quality to the states.

Congress increased the federal role somewhat in the Clean Air Act of 1963. The secretary of the Department of Health, Education, and Welfare was authorized to call abatement conferences when air pollution from one state put citizens of another state in danger, but the Clean Air Act failed to include any sanctions for the enforcement of national standards. Meanwhile, evidence was accumulating that air pollution posed a serious threat to public health throughout the country. Incidents such as the November, 1966, acute air-pollution episode in New York City, an event blamed for the deaths of some 168 people, served as a sobering example of how polluted America's air had become. President Lyndon B. Johnson's Great Society looked to federal regulation as the only effective way to deal with such matters.

The 1967 Air Quality Act authorized the Department of Health, Education, and Welfare to consult with the states to determine air-quality standards in regions of particular concern, and the states were then given a year to formulate a plan to implement the guidelines. Environmentalists were disappointed that Congress still had not provided minimum standards of air quality or effective means for forcing the states to achieve their goals. The most significant aspect of the act was the authorization of some federal enforcement of vehicular emissions standards, with criminal fines of up to $1,000 for each violation of the standards. Relatively weak requirements based on grams of pollutants emitted per mile took effect for new automobiles in 1968.

## 1970 AMENDMENTS

Widespread support of the environmental movement was demonstrated by the enthusiastic response to Earth Day in 1970, and that same year the first report of the U.S. Council on Environmental Quality called on Congress to enact new laws to deal with several problems, including air pollution. Senator Edmund Muskie, a presidential hopeful, was the acknowledged congressional leader in the campaign for tough environmental reform, and he was the chief author of the 1970 clean air bill. President Richard Nixon also supported an aggressive bill. With this bipartisan support, Congress enacted a far-reaching amendment to the 1963 legislation, the Clean Air Act Extension of 1970, which initiated the federal government's regulation of air pollution.

Addressing perceived weaknesses in the existing

law, the landmark 1970 amendments authorized the newly created Environmental Protection Agency (EPA) to establish standards that would be binding on states. Applying a command-and-control approach to regulation, the centerpiece of the legislation was a program for the EPA to determine National Ambient Air Quality Standards (NAAQS) that would define specific levels of air pollution considered harmful to public health. The EPA was also authorized to set emission limits on hazardous pollutants at levels allowing a sufficient margin of safety. Although states might exercise discretion in choosing how to meet the federal standards, they were required to develop state implementation plans (SIPs) that utilized appropriate measures to reach those standards. States could maintain the air-quality-control programs already in place for existing industrial plants while requiring new plants to meet stricter standards based on the best available technology that was economically feasible.

The 1970 legislation stunned American automobile manufacturers by requiring them to curtail their products' emissions of the "big three" pollutants—hydrocarbons, carbon oxides, and nitrogen oxides—by 90 percent within six years. The technology did not exist to allow them to meet the new standards, although it was hoped that new technology could be developed within the specified time. Most members of Congress, few of whom were willing to see the collapse of the automobile industry, understood that it might be necessary to extend the deadline. In fact, the deadlines for meeting the vehicular emission standards turned out to be excessively ambitious, and waivers for the standards were granted in 1971, 1973, 1974, and 1976.

## 1977 AMENDMENTS

With the enthusiastic support of President Jimmy Carter, Congress passed major revisions to the Clean Air Act on August 4, 1977. In addition to making NAAQS more stringent, the amendments required each state to designate "nonattainment" regions

> ## Carter Comments on the Clean Air Act Amendments
>
> *In his statement on signing the Clean Air Act amendments of 1977 into law, President Jimmy Carter explained the value of the new legislation:*
>
> This act is the culmination of a 3-year effort by the Congress to develop legislation which will continue our progress toward meeting our national clean air goals in all parts of the country. The issues involved in amending the Clean Air Act have been difficult and the debate lengthy. However, I believe that the Congress . . . has adopted a sound and comprehensive program for achieving and preserving healthy air in our Nation.
>
> The automobile industry now has a firm timetable for meeting strict, but achievable emission reductions. That industry now knows with certainty what is required and can devote its full-time energies to designing cars which will further our clean air goals while continuing to improve fuel efficiency. This timetable will be enforced.
>
> With this legislation, we can continue to protect our national parks and our major national wilderness areas and national monuments from the degradation of air pollution. Other clean air areas of the country will also be protected, at the same time permitting economic growth in an environmentally sound manner.
>
> The act provides us with a new tool to help abate industrial sources of pollution by authorizing use of economic incentives to reduce noncompliance. By directing the Environmental Protection Agency to establish monetary penalties equal to the cost of cleanup, those industries which delay installing abatement equipment will no longer be rewarded in the marketplace.
>
> These three major provisions, coupled with the other authorities of H.R. 6161, provide the statutory framework for the Environmental Protection Agency to implement a firm, but responsible program for meeting and maintaining air quality standards which are necessary to protect the health of all of our citizens.

based on the NAAQS. Each state was then given the choice of either accepting statutory sanctions or revising its SIPs in order to meet the standards in a timely way. The amendments focused especially on coal-burning power plants, a significant source of sulfur dioxide in the atmosphere, which contributes to acid rain. Existing stationary sources of pollution were required to provide for "reasonably available control technology," and new or modified stationary sources were required to utilize technology meeting the "lowest achievable emission rate," which usually meant the use of expensive scrubbers.

In the case of clean air regions already in attainment, the amendments instituted the Prevention of Significant Deterioration (PSD) program, which was designed to prevent the EPA from allowing deterioration of air quality up to the national standards. Members of Congress from rural districts had unsuccessfully argued that such a program would unfairly restrict industrial growth in areas where air pollution was not a problem.

The 1977 amendments further extended the deadlines by which automobiles were required to achieve emissions-control standards. The stricter controls were scheduled for the 1980 model year. American automakers had insisted that they could not meet the requirements in the existing law, and they had threatened to shut down production lines. This marked the fifth relaxation of the vehicular deadlines.

### 1990 AMENDMENTS

Between 1977 and 1990 numerous efforts were undertaken both to strengthen and to weaken the Clean Air Act, but opposing interest groups prevented major changes in either direction. While there was widespread agreement that the Clean Air Act had been somewhat successful in improving air quality, the administration of President Ronald Reagan strongly opposed any expansion of environmental regulations. During the 1988 presidential election campaign, candidate George H. W. Bush pledged to be the "environmental president." When Bush entered the White House, the deadlines for compliance with most air-quality standards had passed, putting noncompliance regions in danger of losing many industrial jobs. In July, 1989, the Bush administration made a sweeping proposal that most Democrats and environmentalists could support, while conservative Republicans were divided on the issue. The resulting amendments were signed into law on November 15, 1990.

The 1990 amendments were extremely complex, requiring more than seven hundred pages. The major regulatory change, modeled on the Clean Water Act, was a requirement that all major sources of air pollution obtain state operating permits, with the EPA given the authority to veto such permits. The amendments provided additional regulations of emissions that were responsible for acid rain and established an allowance system based on a nationwide limit of 8.1 million metric tons (8.9 million U.S. tons) of sulfur dioxide per year. Other provisions included a phaseout program for chlorofluorocarbons (CFCs) and other ozone-depleting substances, as well as a requirement that industrial plants cut emissions of 189 toxic substances to the levels of the cleanest plants within their particular industries.

Because one important goal of the statute was to

## Time Line of U.S. Clean Air Laws and Policies

| YEAR | EVENT |
| --- | --- |
| 1955 | Air Pollution Control Act, the first U.S. law to address air pollution and fund research into pollution prevention, is passed. |
| 1963 | Clean Air Act is the first U.S. law to provide for the monitoring and control of air pollution. |
| 1967 | Air Quality Act establishes enforcement provisions to reduce interstate air-pollution transport. |
| 1970 | Clean Air Act amendments establish the first comprehensive emission regulatory structure, including the National Ambient Air Quality Standards (NAAQS). |
| 1977 | Clean Air Act amendments provide for the prevention of deterioration in air quality in areas in compliance with the NAAQS. |
| 1990 | Clean Air Act amendments establish programs to control acid precipitation, as well as 189 specific toxic pollutants. |
| 1995 | Oil companies are required to sell reformulated gasoline in metropolitan regions, and gas stations are required to install vapor-retrieval devices on pumps. |
| 2003 | Proposed Clear Skies Bill is designed to amend the Clean Air Act with a cap-and-trade system. |
| 2005 | The EPA's Clean Air Interstate Rule (CAIR) begins a cap-and-trade program to keep air pollution generated in one state from rendering other states noncompliant with air-quality standards. |
| 2008 | A federal appeals court rules that CAIR exceeds the EPA's regulatory authority but later orders temporary reinstatement. |

decrease urban smog, it included strict controls on automobile emissions and mandates for cleaner-burning fuels. Beginning with 1994 automobiles, tail-pipe exhausts were required to contain 60 percent less nitrogen oxide, and emissions-control equipment was required to last ten years. The EPA was authorized to conduct a study to determine whether stricter standards were needed. A pilot program in California required an increasing number of cars and light trucks to run on batteries or nongasoline fuels. Beginning in 1995, oil companies were required to sell only cleaner-burning reformulated gasoline in the smoggiest metropolitan regions, and gasoline stations were mandated to install devices to capture fumes during refueling.

The 1990 amendments considerably strengthened the enforcement provisions under the Clean Air Act. The EPA acquired new powers to issue administrative penalties of up to $25,000 per day, and individuals were empowered to take civil action against polluters. The EPA and the U.S. Department of Justice were given new authority to prosecute misdemeanor and felony violations of the act. The amendments increased maximum sentences for most violations from six months to two years and increased maximum fines from $25,000 to $500,000. An individual who released hazardous air pollutants into the air could henceforth be sentenced to fifteen years in prison and fined up to $250,000, and corporations could be fined up to $1 million.

At an estimated cost of $25 billion per year, the 1990 act is considered the most expensive piece of environmental legislation ever passed. It was expected that the costs would mostly be passed on to consumers in higher prices for cars, gasoline, electricity, and products containing chemicals. The EPA's estimated monetary value of the act, based on its public health and environmental benefits, offsets its cost.

## THE CLEAR SKIES INITIATIVE

In 2002, President George W. Bush announced the Clear Skies Initiative, a policy intended to incentivize innovation and cut costs through a market-based cap-and-trade program for reducing power plant emissions. Through such a program, polluters would have the right to emit a certain quantity of pollutants; a polluter that wishes to emit more than its allowance would have to purchase credits from one that emits less. The initiative, which prioritized economic growth, operated on the assumption that the market drives advances in pollution-control technologies and

thereby hastens environmental progress. The initiative gave rise to the Clear Skies Bill of 2003, which would have amended the Clean Air Act with a cap-and-trade system. Critics—among them the Sierra Club and the Natural Resources Defense Council—argued that the proposed law was a propagandistically named reduction of air-quality protections that would allow increased toxic industrial emissions and hamper enforcement of pollution-control standards. The bill never moved beyond the Senate Environment and Public Works Committee and thus did not become law.

In 2005 the EPA introduced the Clean Air Interstate Rule (CAIR). A key measure of the deadlocked Clear Skies Bill, CAIR is a cap-and-trade program intended to ensure that air pollution generated in one state does not prevent another downwind state from meeting air-quality standards. CAIR, which was designed to reduce smog and soot pollution from power plants in the eastern United States, includes a permanent cap on the precursor pollutants sulfur dioxide and nitrogen oxides. In July, 2008, a federal appeals court vacated CAIR, citing several fundamental flaws in the rule. The EPA, the Environmental Defense Fund, and several states successfully appealed for a rehearing. The court determined that, despite the rule's shortcomings, the environmental and health benefits of CAIR are significant. (The EPA estimated that CAIR would prevent seventeen thousand deaths annually by 2015.) In December, 2008, the court issued an order temporarily reinstating CAIR until the EPA could replace it with a rule that fully addresses CAIR's flaws.

Another measure of the Clear Skies Bill, the Clean Air Mercury Rule (CAMR), was also introduced in 2005. Environmental groups, several states, Native American tribes, and physicians' organizations opposed the rule, as its use of a cap-and-trade program in the case of a bioaccumulative, environmentally persistent material such as mercury would allow the development of toxic hot spots that would endanger human health. CAMR also removed oil- and coal-fired electric-utility steam-generating units from the list of hazardous air-pollutant sources. In 2008, a federal appeals court found CAMR to be in violation of the Clear Air Act and vacated the rule.

## SUBSEQUENT DEVELOPMENTS

In a 2007 case heard by the U.S. Supreme Court, twelve states, three major U.S. cities, a U.S. territory, and several nongovernmental organizations sued the

EPA for failing to regulate greenhouse gases (GHGs) as pollutants. The Court found that the EPA was again in violation of the Clean Air Act and charged the agency with determining whether GHG emissions from new vehicles are pollutants that endanger the public health or welfare. The EPA concluded that GHGs do in fact pose a danger to the public and submitted its endangerment finding to the White House. White House officials refused to read the EPA's report and took other measures to block the EPA's regulation of GHGs during George W. Bush's administration.

In 2009, under President Barack Obama's administration, the EPA issued its final findings regarding GHGs. It determined that current and projected concentrations of six GHGs in the atmosphere—carbon monoxide, methane, nitrous oxide, hydrofluorocarbons, perfluorocarbons, and sulfur hexafluoride—constitute a threat to the public health and welfare of current and future generations. It also found that new motor vehicles were contributing to GHG pollution. These findings may someday lead to more restrictive emissions limits for power plants, oil refineries, auto manufacturers, and other major GHG contributors.

ENFORCEMENT

Based on a command-and-control model, the Clean Air Act and its amendments provide a variety of strong mechanisms for enforcing their statutory and regulatory requirements. The EPA has primary responsibility for enforcement at the federal level, and the states share responsibility for regulating SIPs. Citizens are also given broad opportunities to participate in the enforcement process.

When the EPA finds evidence that a violation has occurred, a regional office of the agency issues a notice of violation to both the source and the state. Based on its investigations, the EPA has the discretion to determine whether further action is necessary. The agency may issue an administrative order requiring a person or institution to comply with the applicable statute or regulation. If the recipient of the order fails to comply, the EPA may enforce the order through a civil action. If there is probable cause that a crime has occurred, the EPA will initiate a criminal prosecution, but the Department of Justice usually takes charge of the legal actions.

In formulating its SIPs, each state is required to include a program of legal enforcement. The states are usually given the opportunity to lead in initiating en-

forcement action if they wish to do so. If a state does not do so, the EPA has the authority to proceed on its own. Any person, moreover, may bring a civil action against an individual or entity alleged to be in violation of the Clean Air Act. If a violation is proved, any monetary awards must be either turned over to the EPA's "penalty fund" or used for "beneficial mitigation projects."

When the landmark amendments of 1970 were passed, proponents of the act tended to be extremely optimistic about the prospects for achieving national air-quality standards without any serious economic costs. The act envisioned full attainment of the standards by 1975, but this expectation turned out to be unrealistic, especially in regard to ozone. In the 1977 amendments, Congress responded to the problem by explicitly recognizing noncompliance regions, which were thereafter required to improve incrementally. It was even more difficult to formulate vehicular emissions standards that were both meaningful and attainable, in part because no one could be certain about the prospects of technological improvement. When automotive technology improved, moreover, no one could be certain about whether Clean Air Act standards were a primary cause.

By the late 1990's, few people denied that the Clean Air Act had helped decrease air pollution and improve the public health. By its nature, however, such legislation does not completely satisfy everyone. Environmental organizations commonly argue that the EPA has not been aggressive enough in its enforcement efforts, while probusiness groups tend to blame the Clean Air Act for forcing American industries to close their doors and move to poor countries with weaker regulatory protections and a greater toleration for dirty air.

*Thomas T. Lewis*
*Updated by Karen N. Kähler*

FURTHER READING

Bryner, Gary C. *Blue Skies, Green Politics: The Clean Air Act of 1990.* Rev. ed. Washington, D.C.: Congressional Quarterly Press, 1995.

Cohen, Richard E. *Washington at Work: Back Rooms and Clean Air.* 2d ed. Boston: Allyn & Bacon, 1995.

Griffin, Roger D. *Principles of Air Quality Management.* 2d ed. Boca Raton, Fla.: CRC Press, 2007.

Lipton, James P., ed. *Clean Air Act: Interpretation and Analysis.* New York: Nova Science, 2006.

Rajan, Sudhir Chella. *The Enigma of Automobility: Demo-*

*cratic Politics and Pollution Control.* Pittsburgh: University of Pittsburgh Press, 1996.

U.S. Environmental Protection Agency. *The Plain English Guide to the Clean Air Act.* Research Triangle Park, N.C.: Office of Air Quality Planning and Standards, 2007.

# Climate change and human health

CATEGORIES: Weather and climate; human health and the environment

SIGNIFICANCE: Changes in climate can affect the numbers of people who die directly as the result of temperature extremes (either cold or hot) or violent weather and can also increase the ranges of certain diseases and other health problems, which can lead to lesser, but sometimes serious, health effects.

Direct deaths from extremely cold or warm weather (hypothermia, heatstroke) are relatively rare in developed countries, but occasionally large numbers of people are killed or otherwise seriously affected by heart attacks caused by the weather (including people trying to shovel too much snow at once). The global warming that has occurred over the past century, and is generally expected to continue, can affect the number of deaths related to weather—reducing the number harmed by severe cold but increasing the number harmed by extreme heat. Climate changes can also alter the numbers of people affected by flood and drought as well as the ranges of parasites and disease vectors.

## DIRECT EFFECTS

Warmer temperatures tend to lead to more frequent deaths from excessive heat, although this is mitigated in the case of greenhouse gas warming by the fact that the greatest warming occurs at night and thus in winter (and especially in the Arctic and Antarctic regions), when the temperatures otherwise would be cooler. Warmer temperatures also reduce the numbers of deaths associated with extreme cold (including cardiovascular and pulmonary diseases, such as influenza). There are actually more deaths from cold than from heat in many areas that face both threats. In Europe, estimated annual deaths amount to approximately 1.5 million from cold compared to about 200,000 from heat; the warmer United States suffered twice as many deaths from cold as from heat from 1979 through 1997. Even the severe heat wave of August, 2003, which led to an estimated 35,000 deaths in Europe (nearly half in France, partly because most doctors who practiced there were away on vacation at the time), resulted in only a modest increase in the number of deaths that year from excessive heat.

Climate change can also alter rainfall patterns, causing some areas to be more subject to floods (which inflict heavy damage and can also lead indirectly to other health problems) or droughts (which can lead to crop failures and water shortages). Scientists do not yet know, however, whether the trend toward warming will result in more floods, more droughts, or even both. (The claim that global warming leads to more extreme weather is still speculative, and disputed by many hurricane specialists.) The Medieval Warm Period (from around 800 to 1250) seems to have led to more droughts overall, but there is no guarantee that the pattern will repeat.

## FOOD AND WATER

Climate changes can also affect food crops. Judging from the past experience of the Medieval Warm Period, this can lead to greater production in some places (partly from longer growing seasons as well as the fertilizing effect of increased carbon dioxide) and shortages in others (caused by droughts and floods rather than the temperature changes, though this can change what crops are grown in particular areas). These shortages can lead to malnutrition, including deficiency diseases, or even famine in poor countries.

Drought also makes obtaining water supplies more difficult even as a population continues to grow. One consequence is that people are often forced to work hard (expending labor that would otherwise be available for other needs) for water that is often tainted, which leads to increasing outbreaks of diseases such as dysentery, typhoid fever, and cholera, as well as aquatic parasites such as guinea worms. When water is scarce, cleaning and other sanitation practices suffer. Unclean bodies (especially hands) help spread diseases, and unclean clothes can carry and spread parasites such as lice.

## DISEASES

Climate changes can affect the ranges of various life-forms in many ways. Warmer weather, particularly if it is also wetter, tends to increase the numbers of insects; the mild winters created by greenhouse gas warming are especially important for those insects that are susceptible to freezing temperatures. Many of these are disease vectors, spreading serious diseases such as malaria, dengue fever, yellow fever, typhus, and the plague. Not only may these insects cover larger areas (and also spread to higher elevations, as shown by a 1997 malaria outbreak in Papua New Guinea at an altitude of 2,100 meters, or 6,900 feet), but warmer temperatures can also enable them to be active for a longer portion of the year. This is especially crucial for mosquitoes that carry dengue fever, but malaria exposure may also increase. Although most malaria victims survive, the disease is very persistent, with frequent relapses, and thus very debilitating. On the other hand, it is estimated that warming will reduce the incidence of schistosomiasis, and possibly also the range of ticks that carry diseases such as Rocky Mountain spotted fever.

In areas that become significantly wetter, increased molds can lead to increases in hay fever and asthma, which can be fatal. Flooding can drive rodents, which help spread diseases such as the plague, from their burrows. When carbon dioxide increases, crop yields improve, but so does the growth of allergenic pollens such as ragweed.

Many skeptics, such as virologist Barry Beaty of Colorado State University, argue that the spread of diseases such as malaria seen in the late twentieth and early twenty-first centuries is the result primarily of nonclimatic factors, such as resistance to drugs by the disease pathogens, resistance to pesticides by the disease vectors, and a collapse in public health measures in some areas. (More than 80 percent of the world population is theoretically vulnerable to malaria even without global warming.) Danish economist Bjørn Lomborg has argued that the most cost-effective way to deal with the various problems resulting from global warming is to fix the individual problems, such as improving public health and medical care, implementing desalination projects and improving infrastructure (pipes and faucets) to supply potable water, and improving the distribution of food.

*Timothy Lane*

FURTHER READING

Braasch, Gary. *Earth Under Fire: How Global Warming Is Changing the World.* Updated ed. Berkeley: University of California Press, 2009.

Fagan, Brian. *The Great Warming: Climate Change and the Rise and Fall of Civilizations.* New York: Bloomsbury, 2008.

Johansen, Bruce E. *The Global Warming Desk Reference.* Westport, Conn.: Greenwood Press, 2002.

Lomborg, Bjørn. *Cool It: The Skeptical Environmentalist's Guide to Global Warming.* New York: Alfred A. Knopf, 2007.

Mann, Michael E., and Lee R. Kump. *Dire Predictions: Understanding Global Warming.* New York: DK, 2008.

Michaels, Patrick J., and Robert C. Balling, Jr. *Climate of Extremes: Global Warming Science They Don't Want You to Know.* Washington, D.C.: Cato Institute, 2009.

Rosenberg, Tina. "The Burden of Thirst." *National Geographic*, April, 2010, 99-115.

Singer, S. Fred, and Dennis T. Avery. *Unstoppable Global Warming: Every 1,500 Years.* Updated ed. Lanham, Md.: Rowman & Littlefield, 2008.

## Climate models

CATEGORY: Weather and climate

DEFINITION: Computer tools that use numerical representations of the climate system to predict future climate

SIGNIFICANCE: Scientific debates and policy recommendations regarding the severity and the causes of changes in the earth's climate are based primarily on the models that scientists use to project climate trends.

Climate models can be used to explore various scenarios; scientists use such models both to study the past (such as linking the ice ages to the rise of the Himalayas) and to project what may occur in the future. No matter how precise or accurately worked a model is, however, the information it provides is merely theoretical and must be verified against actual data, either experimental or (in the case of climate) observational. In addition, a model that accurately reflects the recent known past may have been written to do so.

The key models used in climate research are extremely complex general circulation models (GCMs). An ideal GCM would take into account every factor in climate, but in practice some factors are either ignored or simplified, even in the most complex models. Among these factors are volcanic eruptions, solar output, oscillating weather patterns (some taking a few years, others several decades), the water cycle (evaporation, condensation into low-level clouds, and precipitation all move heat from the surface to the troposphere), atmospheric content (including greenhouse gases such as water vapor and carbon dioxide, as well as pollutants such as sulfur dioxide), ocean currents, wind patterns, and land use (urban areas are much warmer than farmland, which is warmer than forest). None of these is entirely predictable, and some are random in occurrence. Local events can have major global effects; for example, El Niño conditions in the southern Pacific lead to weaker northern Atlantic hurricanes due to wind shear, as well as changes in precipitation in many areas.

The models used by the Intergovernmental Panel on Climate Change generally project a temperature rise of approximately 2 to 3 degrees Celsius (3.6 to 5.4 degrees Fahrenheit) by the year 2100. Part of this rise is expected to come from natural causes, part from increased atmospheric greenhouse gases, and part from positive feedback effects such as increased humidity from evaporation and increased summer snow melt (bare ground absorbs more heat). All of these are speculative numbers. In reality, temperature rise during the late twentieth and early twenty-first century warming period was only approximately one-half of one degree over thirty-five years, less than projected. Part of this may be from the cooling effect of sulfate aerosols (another speculative number), but part may also be from negative feedback effects, such as the water cycle. Scientists continue to disagree regarding the effects of global warming on cyclones and other severe storms; no observational evidence supports the theory that warming leads to more storms or more severe ones. Climate models also make projections about regional conditions (such as increased drought in the southwestern United States) that are not verifiable.

Climate models that look at the effects of greenhouse gases indicate that warming should be far greater in the higher latitudes than near the equator; this is reasonably accurate in the Northern Hemisphere, but not in the Southern Hemisphere owing to the cooling of most of Antarctica. These models project greater warming in the middle troposphere than at the surface, but cooling in the stratosphere; the stratosphere is cooling, but the middle-troposphere

warming is actually less than the surface warming. Modifications made to early 1990's models to account for the effects of sulfate aerosols made the models more accurate overall, but they have failed to explain why there has been less warming than predicted in the Southern Hemisphere (where sulfates in the atmosphere are much smaller). Some of these results are different from those expected with a purely cyclic explanation of global warming (which predicts stratospheric warming, for example).

*Timothy Lane*

FURTHER READING

Mann, Michael E., and Lee R. Kump. *Dire Predictions: Understanding Global Warming.* New York: DK, 2008.

Michaels, Patrick J., and Robert C. Balling, Jr. *Climate of Extremes: Global Warming Science They Don't Want You to Know.* Washington, D.C.: Cato Institute, 2009.

Spencer, Roy W. *Climate Confusion: How Global Warming Hysteria Leads to Bad Science, Pandering Politicians, and Misguided Policies That Hurt the Poor.* New York: Encounter Books, 2008.

# Climatology

CATEGORY: Weather and climate

DEFINITION: The study of factors that produce and influence a region's weather, both on an ongoing basis and by retrospective analysis of historical, archaeological, and geological records

SIGNIFICANCE: The principal aim of climatology is long-range regional weather prediction, traditionally for determining optimum agricultural practices and planning for extreme weather events. With a growing awareness that human activities potentially can have profound negative impacts on climate, policy makers have increasingly turned to climatologists and their models for recommendations on how to shape global energy policies.

Since the dawn of civilization and perhaps even before, human beings have observed regional weather patterns and attempted to correlate them with other phenomena in order to project the patterns into the future, chiefly for agricultural purposes. At low latitudes, where drought is the chief concern, predictions focused on variability in rainfall, while in northerly climates latitudinal variation in temperature and the factors influencing yearly variations in temperature were more important.

## CLIMATOLOGY IN HISTORY

Over the centuries, the perception of human influence over climate has shifted. Ancient civilizations credited ritual observances with the capacity to influence the weather and ascribed extreme weather events such as prolonged droughts to divine wrath at human wrongdoing. With the rise first of modern astronomy, which eliminated the superstition from generally valid astrological long-term weather prediction, and then of increasingly sophisticated worldwide measurement of atmospheric phenomena, educated people came to regard climate as something independent of human activity and not susceptible to modification, except perhaps on a very local or temporary level. The pendulum swung in the opposite direction with the growing recognition that the rapid release of large volumes of greenhouse gases into the atmosphere has produced a general warming trend.

Like meteorology, climatology as a systematic science dates from the middle of the nineteenth century and represents a response to the need by the British Empire and the United States to expand agricultural and economic systems originating in Western Europe to regions with more variable and extreme climates. Close study of droughts and famines in India led to the discovery of the Southern Oscillation (now known as El Niño/Southern Oscillation), a regular decadal fluctuation of pressure in the western Pacific that profoundly affects rainfall in India and elsewhere.

## PALEOCLIMATOLOGY

In order to understand long-term climate fluctuations and trends, climatologists must reconstruct temperatures and rainfall for periods and in geographical areas for which no actual measurements exist—from within the last century in the case of remote and undeveloped regions to hundreds of millions of years in the case of the "snowball earth" of the late Precambrian or the Permian-Triassic extinction event. The climatic events since the last ice age are of most interest in the assessment of the probable effects of human activities and how they might interact with abiotic forcing mechanisms.

On a regional level, vegetation, which can be reconstructed through pollen profiles in sediments or sedimentary rock, is a reliable indicator of climate.

For a picture of global climate, scientists look at ice cores from Greenland and Antarctica, in which trapped air provides a snapshot of atmospheric conditions, including greenhouse gases and particulates, and isotope ratios reflect oceanic evaporation and photosynthesis.

Glacial geologic features provide a persistent legacy of cold conditions, and rapid deposition of deltas and lake sediments is a signature of high rainfall. Where isotope, geologic, and fossil evidence co-occur they present a consistent picture of paleoclimate, but the value of isotope ratios alone is open to question.

## GLOBAL WARMING

Climatology enters into debates on environmental issues only if human actions are perceived as influencing climate. Until fairly recently this would have been mainly in connection with desertification due to overgrazing or forest destruction. Although these human activities undoubtedly play a part in increasing aridity in areas such as the Sahel of Africa, the most recent treatments of the subject place more emphasis on shifts in global weather patterns than on local misuse of resources.

Data gathered and correlated by climatologists firmly identified the trend toward global warming. A network of interconnected national and international agencies tracks the course of global warming in time and supplies the data to bodies concerned with implementing public policy. By monitoring all of the factors believed to influence global temperatures, climatologists can estimate the degree to which warming is caused by human activities and pinpoint which activities produce the largest effects. Despite all of the effort and expertise that go into accumulating and analyzing these data, the results are often inconclusive and open to interpretation; even when unequivocal, the data can become tools for policy makers whose principal aim may not be the preservation of a sustainable environment.

*Martha A. Sherwood*

## FURTHER READING

Alverson, Keith D., Raymond S. Bradley, and Thomas F. Pedersen, eds. *Paleoclimate, Global Change, and the Future.* New York: Springer, 2003.

Bonan, Gordon B. *Ecological Climatology: Concepts and Applications.* New York: Cambridge University Press, 2002.

Leroux, Marcel. *Global Warming: Myth or Reality? The Erring Ways of Climatology.* New York: Springer, 2005.

Mayewsky, Paul, et al. "Holocene Climate Variability." *Quaternary Research* 62 (2004): 243-255.

Oliver, John E., ed. *Encyclopedia of World Climatology.* New York: Springer, 2005.

Saltzman, Barry. *Dynamical Paleoclimatology: Generalized Theory of Global Climate Change.* San Diego, Calif.: Academic Press, 2002.

Thompson, Russell D., and Allen Perry, eds. *Applied Climatology: Principles and Practice.* New York: Routledge, 1997.

# Coal ash storage

CATEGORY: Waste and waste management

DEFINITION: Storage of the waste products of coal burning

SIGNIFICANCE: Recycling the ash that results from coal burning provides a way to remove a waste product. Coal ash contains several environmental toxins, however, and unregulated storage of coal ash allows these toxins to leach into groundwater and pollute local water sources.

Many power plants burn coal to generate energy. The by-products of coal burning are called coal combustion by-products (CCBs) or coal ash. In 2007, coal-burning power plants in the United States generated approximately 131 million tons of coal ash.

Coal ash is generally divided into four types. Fly ash, the finest form of coal ash, forms from non combustible matter in coal. Bottom ash, a coarse and granular material, comes from the bottom of coal-burning furnaces. Coal burned in a cyclone boiler produces a molten ash that, once cooled with water, forms a black material called boiler slag. Gases generated by coal burning, if passed through air-pollution-control systems called scrubbers that remove sulfur, generate flue gas desulfurization (FGD) gypsum.

Recycling and disposal represent the two options for managing coal ash. All four types of coal ash have several beneficial uses. Fly ash can substitute for Portland cement in concrete and produce stronger, less porous, and cheaper forms of concrete. Bottom ash and boiler slag contribute to road base, asphalt paving, and fill material, and engineers have used boiler slag for blasting grit, roofing-shingle aggregate, and snow and ice control. Farmers use FGD gypsum as a soil amendment, and builders use it to make wallboard. About one-third of coal ash and one-fourth of scrubber waste are recycled.

Methods of disposing of coal ash are divided into dry disposal, in which unwanted CCBs are transported to landfills, and wet disposal, in which coal ash is sluiced to storage lagoons. Combined coal ash in storage lagoons, known as ponded ash, accounts for approximately 30 percent of all disposed coal ash.

Coal ash often contains poisonous materials, such as arsenic, cadmium, chromium, lead, selenium, and other toxins that can cause cancer, liver damage, and neurological problems in humans and can kill fish and other wildlife. Toxins from coal ash storage lagoons can leach into groundwater. Studies by the U.S. Environmental Protection Agency (EPA) have shown that coal ash dumps can significantly increase the concentration of toxic metals in local drinking water. Worse still, coal ash used as fill-in material can ooze poisons into groundwater.

The retaining walls that enclose coal ash storage lagoons can also fail and produce environmental disasters. For example, on December 22, 2008, the ash dike enclosing a containment pond at the Tennessee Valley Authority's Kingston Fossil coal-burning plant, near Harriman, Tennessee, broke. Approximately one billion gallons of coal fly ash slurry flowed into tributaries of the Tennessee River, which supplies the drinking water for Chattanooga, Tennessee, and those who live downstream in Alabama, Tennessee, and Kentucky.

Because coal ash can be reused, it is not classified as a hazardous waste in the United States, nor is its reuse subject to federal oversight. The EPA moved to close this loophole in 2009, but the draft rule was held up by the Office of Management and Budget. The EPA's stated purpose was to prevent environmental damage from coal ash ponds by requiring such ponds to have synthetic liners and leachate collections systems and by phasing out leak-prone ash ponds.

*Michael A. Buratovich*

FURTHER READING

Freese, Barbara. *Coal: A Human History.* Cambridge, Mass.: Perseus, 2003.

Goodell, Jeff. *Big Coal: The Dirty Secret Behind America's Energy Future.* New York: Houghton Mifflin, 2006.

Greb, Stephen F., et al. *Coal and the Environment.* Alexandria, Va.: American Geological Institute, 2006.

# Convention on Long-Range Transboundary Air Pollution

CATEGORIES: Treaties, laws, and court cases; atmosphere and air pollution

THE CONVENTION: International agreement to limit air pollution, including pollution created in one country that affects the environment in another

DATE: Opened for signature on November 13, 1979

SIGNIFICANCE: Although the Convention on Long-Range Transboundary Air Pollution has had little direct effect on air quality, it is important because it was the first agreement among nations of Eastern Europe, Western Europe, and North America regarding the environment.

Until the 1970's most local, regional, and national regulations regarding industrial air pollution were concerned only with pollution generated in the immediate area. For example, regulations in a particular community might call for taller industrial smokestacks to carry pollution farther away, but there was little official concern about where that pollution might eventually return to earth. Similarly, local assessments and treatments of pollution tended not to consider pollution that might come to an area from distant generators. The only exceptions were a small number of treaties between two countries, such as between the United States and Canada or between Germany and France.

In 1972 the United Nations Conference on the Human Environment, held in Stockholm, Sweden, drew attention to the harmful effects of acid rain, including damage to forests, crops, surface water, and buildings and monuments, especially in Europe. Data revealed that while all European nations were producing alarming levels of air pollution, several nations were receiving more pollution from beyond their borders than they were generating on their own. It became clear that pollution is both imported and exported, that sulfur and nitrogen compounds can travel through the air for thousands of miles, and that any serious attempt to deal with air pollution must reach beyond political boundaries. Two major studies of the long-range transport of air pollutants (LRTAP), conducted under United Nations sponsorship in 1972 and 1977, conclusively proved that air pollution is an international—even a global—problem caused primarily by fossil-fuel combustion and harming both industrial and nonin-

dustrial nations around the world.

In 1979 the United Nations Environment Programme organized a convention in Geneva, Switzerland, for the thirty-four member countries of the United Nations Economic Commission for Europe (UNECE), a group that includes all European nations, the United States, and Canada. Significantly, the gathering had the participation of Eastern European nations under the Soviet Union, marking the first time these nations had collaborated with Western Europe to solve an international environmental problem. The Convention on Long-Range Transboundary Air Pollution was signed by thirty-two nations on November 13, 1979, and went into effect on March 16, 1983. It called upon signatory nations to limit and eventually reduce air pollution, in particular sulfur emissions, using the best and most economically feasible technology; to share scientific and technical information regarding air pollution and its reduction; to permit transboundary monitoring; and to collaborate in developing new antipollution policies. Under the terms of the convention, an international panel would undertake a comprehensive review every four years to determine whether goals were being met, and an executive body would meet each year.

The convention did not include any specific plan for the reduction of air pollution. It did not contain any language calling for particular amounts by which emissions would be reduced, nor did it include a schedule by which the reductions would occur. Scandinavian nations, which were among the countries most affected by acid rain, urged the other participants to adopt these kinds of policies, but other countries, led by the United States, the United Kingdom, and West Germany, defeated the proposal.

In the years after the convention went into force, however, several nations did make commitments to reduce emissions by specific amounts, including West Germany, which changed its position as further information was revealed about deforestation caused by acid rain. At the 1983 executive body meeting, eight nations, including Canada, West Germany, and the Scandinavian countries, made a formal commitment to reduce their emissions by 30 percent by 1993, using 1980 levels as the baseline. Over the next two years thirteen more nations announced similar goals, and in 1985 the commitment to a 30 percent reduction was formally adopted as an amendment to the convention that was signed by nineteen nations.

Neither the United States nor the United Kingdom

## Fundamental Principles of the Convention on Long-Range Transboundary Air Pollution

*Articles 2 through 5 of the Convention on Long-Range Transboundary Air Pollution set out the fundamental principles of the agreement, by which all contracting parties agree to be bound.*

ARTICLE 2:

The Contracting Parties, taking due account of the facts and problems involved, are determined to protect man and his environment against air pollution and shall endeavour to limit and, as far as possible, gradually reduce and prevent air pollution including long-range transboundary air pollution.

ARTICLE 3:

The Contracting Parties, within the framework of the present Convention, shall by means of exchanges of information, consultation, research and monitoring, develop without undue delay policies and strategies which shall serve as a means of combating the discharge of air pollutants, taking into account efforts already made at national and international levels.

ARTICLE 4:

The Contracting Parties shall exchange information on and review their policies, scientific activities and technical measures aimed at combating, as far as possible, the discharge of air pollutants which may have adverse effects, thereby contributing to the reduction of air pollution including long-range transboundary air pollution.

ARTICLE 5:

Consultations shall be held, upon request, at an early stage between, on the one hand, Contracting Parties which are actually affected by or exposed to a significant risk of long-range transboundary air pollution and, on the other hand, Contracting Parties within which and subject to whose jurisdiction a significant contribution to long-range transboundary air pollution originates, or could originate, in connection with activities carried on or contemplated therein.

---

agreed to the 30 percent reductions, and neither country signed the 1985 protocol. The United Kingdom informally agreed to attempt to reduce emissions by 30 percent but was unwilling to commit the financial resources to guarantee the reduction, especially since the benefits were uncertain. In fact, many scientists felt that a 30 percent reduction would not be enough to yield significant improvement. The United States argued that it had already taken major steps to reduce its emissions prior to 1980, so using 1980 data as a baseline would subject the United States to unrealistic and unfair demands for further reduction. This refusal to ratify the protocol caused tension between the United States and Canada, because much of the air pollution that affects eastern Canada comes from the Great Lakes industrial belt in the United States.

*Cynthia A. Bily*

FURTHER READING

Brunnee, Jutta. *Acid Rain and Ozone Layer Depletion: International Law and Regulation.* Boston: Hotei, 1988.

Elsom, Derek M. *Atmospheric Pollution: Causes, Effects, and Control Policies.* Malden, Mass.: Blackwell, 1987.

Fishman, Jack. *Global Alert: The Ozone Pollution Crisis.* New York: Plenum, 1990.

Sand, Peter H. "Air Pollution in Europe: International Policy Responses." *Environment* 29 (December 1987): 16-20, 28-29.

Visgilio, Gerald R., and Diana M. Whitelaw, eds. *Acid in the Environment: Lessons Learned and Future Prospects.* New York: Springer, 2007.

Wetstone, G. S., and A. Rosencranz. *Acid Rain in Europe and North America: National Responses to an International Problem.* Washington, D.C.: Environmental Law Institute, 1983.

## Detoxification

CATEGORY: Human health and the environment

DEFINITION: Reduction or elimination of the toxic properties of a substance to make it less harmful to or more compatible with the environment

SIGNIFICANCE: Hazardous substances often enter the environment as the result of various manufacturing processes and other human activities. The detoxification of air, water, and soil that have been negatively affected by such substances can help to minimize environmental damage.

Increasing industrialization during the twentieth century led to the release of large amounts of hazardous waste and by-products into the environment. Pesticides were another source of toxins, as agriculture worked to maintain the crop yields necessary to feed the growing population of the world. Some toxins are analogues of harmful substances that occur naturally and may degrade rapidly by natural means. Others are more persistent in the environment and produce unwanted effects. "Detoxification" is the general term applied to the various processes by which toxins are removed from the environment or are rendered less harmful.

## Toxic Substances

A substance is considered hazardous if it poses a threat to human health or the environment when it is spread, treated, disposed of, or transported. Toxic and hazardous substances often occur as a result of the manufacture of materials designed to protect humans and improve quality of life. Sources of hazardous waste include the manufacture of chemicals and allied products, the manufacture of petroleum and coal products, the primary metals industry, and the metals fabrication industry. Environmental releases of toxic chemicals may occur unintentionally through emissions from compressors, pump seals, valves, spills, pipelines, and storage tanks, or intentionally through discharges of wastes into air or water or through inappropriate disposal in landfills. The Environmental Protection Agency (EPA) has reported that Americans generate 1.6 million tons of household hazardous wastes each year.

The disposal of hazardous substances is not a simple matter. Many toxic substances are not suitable for disposal in regular landfills used for trash. Some hazardous substances are water-soluble and can leach through the soil into rivers, lakes, and groundwater supplies to pollute sources of potable water. Some wastes have a significant vapor pressure and can be spread over wide areas by wind and air currents. Corrosive wastes must be disposed of in containers that will not decompose.

Public concerns regarding toxic substances in the environment have elicited different approaches to solving these problems. Environmental activists have advocated the use of natural pesticides and nonpolluting agricultural chemicals. The U.S. Congress has addressed the issue of toxic substances in the environment with regulations that specify detoxification procedures for wastewater, contaminated soil, and landfills. These regulations include the Federal Water Pollution Control Act (1974), the Safe Drinking Water Act (1974), and the Federal Environmental Pesticide Control Act (1972).

## Methods of Detoxification

Many natural processes cause detoxification of harmful substances in the environment. Gaseous pollutants or toxins that are exposed to sunlight are subject to photochemical decomposition, in which ultraviolet light causes bonds within the compounds to break. The resulting fragments react with oxygen (oxidation) or water (hydration) to form less toxic compounds. These may undergo repeated degradation in the same manner. Microbial degradation, in which organisms metabolize a wide variety of organic compounds to carbon dioxide and water or convert them into less harmful substances, promotes detoxification of many organic toxins. Some newer pesticides, such as organophosphates, are designed to degrade on repeated exposure to water, forming relatively harmless products. Earlier pesticides, such as polychlorinated biphenyls (PCBs), were found to degrade slowly in the environment. Toxins with slow detoxification pathways bioaccumulate in organisms, causing harmful effects on fish and wildlife. Such effects may be magnified in the food chain.

Efforts to supplement natural detoxification processes include enzymatic (biological) and other chemical methods. Many microorganisms capable of metabolizing toxins have been isolated and cultured in order to treat hazardous wastes. Such treatments are usually carried out at regional waste management centers. One type of chemical treatment involves chelation or precipitation. This is useful for eliminating metals, either in ionic or elemental form, from water and soil. In this method, an organic compound forms an insoluble precipitate with the metal. Filtration removes the precipitate, which can then be subjected to further disposal methods in concentrated form. Composting, or land farming, involves spreading waste materials over a large land area, where they decompose. Pesticides and wastes from paper mills have been detoxified this way. Land farming requires monitoring to ensure that toxins in the wastes do not leach into groundwater.

Thermal treatment is considered a safer process. An example of this type of detoxification method is incineration, during which high temperatures oxidize the solid and liquid organic wastes to carbon dioxide and water in the presence of oxygen. However, people living in communities near incinerators often fear ill effects from possible emissions or leaks. One solution to this concern is the incineration of wastes on ships. *Vulcanus*, a Dutch ship, was used to incinerate large quantities of Agent Orange, a hazardous herbicide contaminated with toxic dioxins.

Another method of detoxification is vitrification, in which toxic materials are converted to glass. Vitrification has been used to dispose of asbestos, which is considered to be a highly hazardous material. It has been reported that vitrification can work with almost any kind of waste, including industrial sludges, soil contaminated by lead, and medical wastes.

*Beth Ann Parker and Massimo D. Bezoari*

FURTHER READING

Brooks, Adrienne C. "A Glass Melange: New Options for Hazardous Wastes." *Science News* 147 (January 21, 1995).

Häggblom, Max M., and Ingeborg D. Bossert, eds. *Dehalogenation: Microbial Processes and Environmental Applications.* Norwell, Mass.: Kluwer Academic, 2003.

Lave, Lester B., and Arthur C. Upton. *Toxic Chemicals, Health, and the Environment.* Baltimore: The Johns Hopkins University Press, 1987.

Newman, Michael C. *Fundamentals of Ecotoxicology.* 3d ed. Boca Raton, Fla.: CRC Press, 2010.

Shen, Samuel K., and Patrick F. Dowd. "Detoxifying Enzymes and Insect Symbionts." *Journal of Chemical Education* 69 (October, 1992): 796-780.

# Dichloro-diphenyl-trichloroethane

CATEGORY: Pollutants and toxins

DEFINITION: Synthetic organochlorine insecticide

SIGNIFICANCE: Dichloro-diphenyl-trichloroethane, better known as DDT, has been used extensively in agriculture and for control of insect-borne diseases worldwide. However, its persistence in the environment and ability to accumulate in the food chain have resulted in devastating consequences to wildlife. The harmful effects of DDT became a major focus for the emerging environmental movement during the 1960's.

During the 1930's scientists began searching for organic (carbon-based) insecticides. Prior to that time, insecticides were mainly derived from toxic metals, such as arsenic and mercury. In 1939, while experimenting with chlorinated hydrocarbons, Swiss chemist Paul Hermann Müller discovered the insecticidal properties of DDT, a chemical that had first been synthesized more than half a century earlier. His findings led to the development of the first synthetic, organic insecticide, which was introduced commercially by the Swiss chemical company J. R. Geigy A.G. in 1942. DDT was initially used to provide protection against typhus to civilians and Allied troops during World War II by killing body lice. Before that, pyrethrum powder was a common means for combating body lice; however, Japan was the chief exporter of this chrysanthemum-derived repellent, and hostilities had left the Allies with insufficient pyrethrum supplies. DDT's key role in suppressing a typhus epidemic in Italy in 1943 led to Müller's receiving the Nobel Prize in Physiology or Medicine in 1948.

When news of DDT's effectiveness was released by the British government in 1944, the U.S. Department of Agriculture (USDA) concluded that before DDT could be recommended for use by farmers, more information about its toxicity was needed. By the 1946 crop year, limited use was permitted even though evidence suggested that DDT might have some acute toxic effects on birds and that it could be stored in animal fat and excreted in milk. Commercial demand for DDT was fueled in large part by accounts of how well it had performed during wartime. The success of DDT served as an impetus for chemical companies to begin an intensive search for other organic pesticides.

Between 1940 and 1980, at least 1.8 billion kilograms (4 billion pounds) of DDT were used. More than 1,200 different formulations were developed for industrial, agricultural, and public health applications in the United States alone. Annual worldwide production peaked in 1964 at 90 million kilograms (198 million pounds).

A NEW POLLUTANT

The insecticidal properties of DDT are related to its ability to act as a nerve poison and to pass freely through insect cuticles. In addition to causing convul-

## Milestones in DDT History

| YEAR | EVENT |
|---|---|
| 1874 | The first synthesis of DDT is reported. |
| 1939 | Paul Müller discovers DDT's insecticidal properties. |
| 1942 | The first commercial DDT formulations are introduced by the Swiss company J. R. Geigy. |
| 1943-1945 | DDT is used on civilians and military troops in Europe for the control of lice and typhus. |
| 1946 | The limited use of DDT on crops is permitted by the U.S. Department of Agriculture. |
| 1948 | Müller receives the Nobel Prize in Physiology or Medicine for the development of DDT as an insecticide. The first insects to develop resistance to DDT are observed. |
| 1950's | DDT is used widely for agriculture, public health, and domestic pest control. Laboratory and field studies reveal the negative effects of DDT. |
| 1957 | The Clear Lake study shows the bioaccumulation of DDT in aquatic life and birds; citizens on Long Island, New York, file a suit in an attempt to halt aerial DDT spraying. |
| 1958 | Robert Barker publishes the results of studies that link DDT to declines in robin populations. |
| 1961 | Annual production levels of DDT in the United States peak at 160 million pounds. |
| 1962 | Rachel Carson publishes *Silent Spring*, which explains the dangers of DDT to a broad audience. |
| 1963 | The President's Science Advisory Committee releases a report on pesticide use that becomes the keystone of the drive to ban DDT. |
| 1964 | The U.S. Federal Commission on Pest Control is established. |
| 1967 | The Environmental Defense Fund (EDF) is formed. |
| 1968 | Joseph Hickey and Daniel Anderson publish a report on DDT's impact on declining raptor populations. |
| 1968 | The Wisconsin Hearings, the first major legal challenge to the use of DDT, begin. |
| 1969 | Malaria is virtually eliminated in China, largely as a result of DDT use. |
| 1969 | Michigan and Arizona become the first states to ban DDT use. |
| 1969 | The EDF files petitions with U.S. federal agencies seeking the elimination of the use of DDT. |
| 1969 | The use of DDT in residential areas is banned in the United States. |
| 1970 | The Environmental Protection Agency (EPA) is established. |
| 1972 | The EPA bans DDT use in the United States. |
| 1990's | Studies suggest that DDT acts as an endocrine disrupter. |
| 1993 | Reports in the *Journal of the National Cancer Institute* claim that DDT may increase the risk of breast cancer. |
| 1998 | International negotiations to phase out the production and use of DDT and other persistent organic pollutants begin in Montreal, Canada. |
| 2000 | South Africa reintroduces the use of DDT to combat the spread of malaria by mosquitoes. |
| 2004 | DDT is among the chemicals covered by the Stockholm Convention on Persistent Organic Pollutants, an international treaty designed to eliminate or reduce the release of toxic, bioaccumulative chemicals. |
| 2008 | Twelve nations, including India and several Southern African countries, are reported to be using DDT to combat malaria. |

sions, paralysis, and death, DDT can also interfere with calcium-dependent processes. Because DDT in crystalline form is not readily absorbed through animal skin (unlike DDT mixed in solution), the compound was initially regarded as a safe alternative to metal-based insecticides.

Although effective, DDT does have undesirable characteristics. As a broad-spectrum insecticide, DDT kills a wide variety of organisms, including beneficial insects such as bees. Also, development of resistance to DDT among pest insects was observed as early as 1948. Because of its chemical composition, DDT is preferentially stored in animal fat and is therefore not readily excreted by animals that ingest it. This fat solubility and DDT's persistence in the environment cause the pesticide to accumulate in the food chain.

The first clear evidence of the bioaccumulation of DDT came from a case study in Clear Lake, California. Between 1949 and 1957 DDT was used to control gnats on the lake. By the mid-1950's, the health of fish-eating birds in the area began to decline; several bird species, especially grebes, were dying in large numbers. Because no infectious agent was found, scientists used new analytical methods developed to measure compounds in tissues. High levels of DDT were detected in plankton, fish, and birds in and around Clear Lake. The studies also clearly showed biomagnification: Levels of pesticide residues were found to be sequentially higher at each step in the food chain, with concentrations in grebes and gulls up to 100,000 times greater than in the formulations of DDT that were sprayed.

By the mid-1950's DDT's toxicity was becoming evident throughout the United States. DDT was used in the Midwest and New England to control the elm bark beetle, an insect that spreads the fungus that causes Dutch elm disease. Several studies between 1954 and 1958 noted sharp declines in robin populations—in some areas by as much as 70 to 90 percent. Extensive aerial spraying for gypsy moths during the 1950's from Michigan to New England coincided with significant declines in many species of songbirds and bees. Ironically, this affected populations of some of the natural predators of the intended target pests.

The DDT spraying also had a negative impact on agriculture. In addition to reduced pollination caused by the loss of bees, farmers were discovering that cows' milk and farm produce were contaminated with pesticide residues. In the Pacific Northwest, DDT used to control the spruce budworm devastated salmon populations. Coastal spraying along the Atlantic Ocean to control the salt marsh mosquito took a heavy toll on migrating birds, marine life, and raptors.

The effect of DDT on raptor populations is well known. In 1968 an article in *Science* magazine by wildlife ecologists Joseph Hickey and Daniel Anderson reported that the decline in populations of birds of prey was largely caused by eggshell breakage caused by chlorinated hydrocarbons. Calcium processes were altered in birds containing high DDT levels in their fatty tissues, resulting in the production of eggs with dangerously thin shells. The young did not hatch because the eggs were crushed during incubation. The American eagle and the osprey ended up on the brink of extinction largely as a result of widespread DDT use.

BAN ON DDT

During the late 1950's the first DDT-related lawsuits were filed over losses to farmers and beekeepers and in attempts to stop further aerial spraying. The most notable case of the time was filed in 1957 by a group of citizens led by well-known ornithologist Robert Cushman Murphy in order to gain an injunction to stop the spraying of DDT over Long Island, New York. The injunction was not granted, but the case went all the way to the U.S. Supreme Court, which declined to hear it.

Perhaps one of the most significant events leading to the ban of DDT was the publication of Rachel Carson's book *Silent Spring* in 1962. The author described the negative environmental impact of pesticides such as DDT, and the subsequent public outcry led to a dramatic decline in DDT use. Production of DDT in the United States peaked in 1961, and global production began to decline around 1964. As a result of the controversy spawned by Carson's book, the President's Science Advisory Committee was charged with reviewing pesticide use. The committee's report, published in 1963, called for legislative measures to safeguard the health of the land and people against pesticides. The Federal Commission on Pest Control was established in 1964, and four governmental committees studied DDT in depth between 1963 and 1969. Ultimately, these investigations led to the establishment of the Environmental Protection Agency (EPA) in 1970.

DDT was also a major impetus for the formation of associations whose missions were aimed at protecting the environment and public health. The newly formed

Environmental Defense Fund (EDF) initiated a series of court hearings and lawsuits related to DDT during the late 1960's. In October, 1969, it filed petitions with the USDA and the Department of Health, Education, and Welfare seeking elimination of the use of DDT. When no effective action resulted, EDF, along with other environmental groups and individuals, took the case to court. On May 28, 1970, the U.S. Court of Appeals for the District of Columbia rendered two major rulings on DDT in response to EDF's litigation. In addition to leading to the eventual ban of DDT, these rulings set important environmental law precedents: They provided power to membership associations (environmental groups) and served to protect public interests.

In 1972 the EPA banned the use of DDT in the United States except in cases of pest-control emergencies (for example, to avoid outbreaks of typhus, bubonic plague, and rabies) and highly restricted the use of other chlorinated hydrocarbons. However, the ban applied only to DDT use within U.S. borders; it still allowed American companies to produce DDT for export, which they continued to do for several years. Many other countries also banned or severely restricted DDT manufacture and use.

PERSISTENCE OF DDT

Even decades after the ban on DDT in the United States and several other countries, the pesticide continues to be found in significant concentrations in marine animals and other wildlife. DDT can be detected in the tissues of almost every person on earth, especially indigenous peoples living in the Arctic and workers from insecticide production plants and agriculture. DDT is present in human breast milk, and it can pass through the placenta from mother to fetus to impair brain development and increase the risk of birth defects.

New concerns about DDT's toxicity arose as a result of studies published beginning during the early 1990's. Data suggest that DDT and its metabolites can act as endocrine disrupters—compounds that mimic naturally occurring hormones in animals. Evidence indicates that such compounds can decrease sperm count and fertility, affect the onset of puberty, alter male and female characteristics in wildlife, increase the risk of cancer of reproductive organs, and otherwise affect growth, development, metabolism, and reproduction.

In 2004 an international treaty restricting pro-

## DDT Ban Takes Effect

*On December 31, 1972, the Environmental Protection Agency issued the following press release regarding the ban on DDT.*

The general use of the pesticide DDT will no longer be legal in the United States after today, ending nearly three decades of application during which time the once-popular chemical was used to control insect pests on crop and forest lands, around homes and gardens, and for industrial and commercial purposes.

An end to the continued domestic usage of the pesticide was decreed on June 14, 1972, when William D. Ruckelshaus, Administrator of the Environmental Protection Agency, issued an order finally cancelling nearly all remaining Federal registrations of DDT products. Public health, quarantine, and a few minor crop uses were excepted, as well as export of the material.

The effective date of the EPA June cancellation action was delayed until the end of this year to permit an orderly transition to substitute pesticides, including the joint development with the U.S. Department of Agriculture of a special program to instruct farmers on safe use of substitutes.

The cancellation decision culminated three years of intensive governmental inquiries into the uses of DDT. As a result of this examination, Ruckelshaus said he was convinced that the continued massive use of DDT posed unacceptable risks to the environment and potential harm to human health.

Major legal challenges to the EPA cancellation of DDT are now pending before the U.S. Court of Appeals for the District of Columbia and the Federal District Court for the Northern District of Mississippi. The courts have not ruled as yet in either of these suits brought by pesticide manufacturers.

DDT was developed as the first of the modern insecticides early in World War II. It was initially used with great effect to combat malaria, typhus, and the other insect-borne human diseases among both military and civilian populations.

A persistent, broad-spectrum compound often termed the "miracle" pesticide, DDT came into wide agricultural and commercial usage in this country in the late 1940s. During the past 30 years, approximately 675,000 tons have been applied domestically. The peak year for use in the United States was 1959 when nearly 80 million pounds were applied. From that high point, usage declined steadily to about 13 million pounds in 1971, most of it applied to cotton.

The decline was attributed to a number of factors including increased insect resistance, development of more effective alternative pesticides, growing public and user concern over adverse environmental side effects—and governmental restriction on DDT use since 1969.

duction and use of DDT entered into force. This treaty, the Stockholm Convention on Persistent Organic Pollutants, is intended to eliminate or reduce the release of several toxic, bioaccumulative chemicals. DDT is among the initial twelve persistent organic pollutants (POPs) specified in the treaty. Unlike most of the other POPs, DDT may be used for disease vector control under the Stockholm Convention until other effective and affordable control methods are available. The convention allows the spraying of indoor walls with DDT, notably as a weapon against malaria and other mosquito-borne tropical diseases, and many countries in Africa and Asia employ DDT in this way. South Africa credits much of its success in controlling malaria in the early twenty-first century to its reintroduction of DDT use in 2000. The World Health Organization recommends that if indoor DDT spraying for mosquito control is conducted, it should be part of an integrated vector management approach—that is, one that uses insecticide-treated mosquito nets, drainage of mosquito-breeding bodies of water, and other methods to discourage overdependence on DDT and development of DDT-resistant mosquito species.

*Diane White Husic*
*Updated by Karen N. Kähler*

FURTHER READING

Carson, Rachel. *Silent Spring.* 40th anniversary ed. Boston: Houghton Mifflin, 2002.

Colborn, Theo, Dianne Dumanoski, and John P. Myers. *Our Stolen Future: Are We Threatening Our Fertility, Intelligence, and Survival?* New York: Plume, 1996.

Dunlap, Thomas R., ed. *DDT, "Silent Spring," and the Rise of Environmentalism: Classic Texts.* Seattle: University of Washington Press, 2008.

Glausiusz, Josie. "Can A Maligned Pesticide Save Lives?" *Discover,* November, 2007, 34-36.

Karasov, William H., and Carlos Martínez del Rio. *Physiological Ecology: How Animals Process Energy, Nutrients, and Toxins.* Princeton, N.J.: Princeton University Press, 2007.

Schapira, Allan. "DDT: A Polluted Debate in Malaria Control." *The Lancet* 368, no. 9553 (2006): 2111-2113.

World Health Organization. *DDT and Its Derivatives: Environmental Aspects.* Geneva: Author, 1989.

# Diquat

CATEGORY: Pollutants and toxins
DEFINITION: Contact herbicide
SIGNIFICANCE: As is the case for many herbicides, diquat has the potential to contaminate supplies of drinking water if it is released into wastewater and subsequently absorbed into soil.

The herbicide diquat, which has been used in the United States since the 1950's, is potentially toxic if swallowed or inhaled. It is not produced in the United States, but nearly one million pounds of the compound are imported each year. Approximately two-thirds of this amount is utilized as desiccant or defoliant, with another one-third used for aquatic weed control. Diquat is readily adsorbed into clay particles in the soil, sediment in water, and the surfaces of weeds. While it is biodegradable, if it is adsorbed into plant life and subject to photodegradation by sunlight, when bound to sediment diquat remains stable for weeks or even months. Ultimately, however, the herbicide undergoes degradation through the action of soil flora.

The extensive overgrowth of weeds affecting both crops and waterways has created significant problems in management. Regulations pertaining to use of herbicides remain under the auspices of the Environmental Protection Agency (EPA) and its various offices. Cutbacks in funding, however, have hampered research into the short- and long-term effects of using herbicides in the control of weed problems; therefore, the long-term effects of diquat are unclear.

Since diquat is a nonselective herbicide, its use is limited. It is used as a growth regulator in suppressing the flowering of sugarcane and to control aquatic weeds in the absence of endangered plant species. However, the most important use of diquat is in the desiccation of potato haulm or seed crops such as clover and alfalfa. Such practices are generally carried out to prepare crops for harvest. Application of diquat results in loss of moisture from the leaf, usually killing the plant. Desiccation may also be utilized in prevention of seed loss resulting from scattering of seed upon opening of the pod. The application of diquat prior to harvest of crops such as alfalfa greatly reduces seed loss.

The primary metabolic effect of diquat seems to be its ability to divert electrons activated during photosynthesis into the production of toxic compounds such as hydrogen peroxide. Peroxide production in turn results in membrane damage to those parts of the plant in contact with the herbicide. Human exposure to diquat is primarily occupational, with agricultural workers who use the chemical at highest risk. Most actual poisonings have been intentional, with diquat used as a means of suicide. Nevertheless, though diquat is not as toxic as some herbicides, the level of its toxicity is mainly a function of degree. Exposure to significant levels of the chemical may cause damage to the central nervous system and kidneys, while cataract formation is the most common effect of chronic exposure.

*Richard Adler*

FURTHER READING

Manahan, Stanley E. "Water Pollution." In *Fundamentals of Environmental Chemistry.* 2d ed. Boca Raton, Fla.: CRC Press, 2001.

Udeh, Patrick J. "Organic Contaminants in Drinking Water." In *A Guide to Healthy Drinking Water: All You Need to Know About the Water You Drink.* Lincoln, Nebr.: iUniverse, 2004.

# Donora, Pennsylvania, temperature inversion

CATEGORIES: Disasters; atmosphere and air pollution

THE EVENT: Temperature inversion resulting in the first outdoor air-pollution disaster in U.S. history

DATES: October 26-31, 1948

SIGNIFICANCE: The temperature inversion event that took place in Donora helped to prompt an increase in research concerning air pollution and raised awareness of the importance of the association between certain meteorological conditions and air-pollution episodes.

Donora is an industrial town of 14,000 located 40 kilometers (25 miles) southeast of Pittsburgh, Pennsylvania, in the Monongahela River Valley. It is surrounded by hills that are 107 meters (350 feet) in elevation. In 1948 four industries with 6-meter (20-foot) smokestacks were located along a 4.8-kilometer (3-mile) stretch of the Monongahela's banks: a steel mill, which used high-sulfur coke; a zinc smelting plant, which used high-sulfur ores; a wire manufacturing plant; and a sulfuric acid plant. In addition, high-sulfur coal was used to generate the area's electricity and to heat most homes and businesses. Therefore, sulfur dioxide fumes and hydrocarbon particulates were common in the air of Donora.

On the morning of Tuesday, October 26, 1948, a surface high-pressure weather cell with light winds settled over the eastern United States. A blocking ridge in the upper air and the westward position of the jet stream prevented any other weather systems from moving into the area. That night, the cool, heavy air from the mountain slopes drained into Donora's valley, and this air combined with the sinking air of the high-pressure cell to create a strong surface temperature inversion.

A temperature inversion occurs whenever the temperature of the atmosphere increases with height. Because cool air is heavier than warm air, a temperature inversion discourages surface air from rising and dissipating surface air pollutants into the upper air. The air within Donora's valley, which was filled with the pollutants being emitted by industries, shops, and homes, became trapped at the surface with no way of dispersing either vertically, because of the temperature inversion, or horizontally, because of the surrounding mountains. To complicate matters, radiation fog, which had begun to form during the afternoon, blocked sunlight from reaching the surface of the earth, preventing surface heating from breaking the inversion.

This meteorological situation remained over Donora for five days. The result was an unprecedented air-pollution episode. By Wednesday morning, visibility in Donora was so poor that local residents were becoming lost. On Friday, respiratory-related health complaints from residents began to pour in; four thousand people became ill from the polluted air between 6:00 A.M. and midnight. The first death occurred early Saturday morning, with sixteen more dead and two thousand more ill by midnight. Three more deaths occurred on Sunday even though relief came as an approaching front pushed the high-pressure cell out of the area and broke the temperature inversion. Subsequent rain washed most of the pollutants out of the air, the wind swept the smoke away, and the disaster was over.

Active responses to the lessons learned from this event were slow to materialize, but the experience in

Donora did eventually prompt an increase in air-pollution research. It also caused many people to question the acceptance of billowing smokestacks as a necessary part of economic progress and to recognize the importance of the association between certain meteorological conditions and air-pollution episodes.

*Kay R. S. Williams*

FURTHER READING

Magoc, Chris J. "The Donora Disaster and the Problem of Air Pollution." In *Environmental Issues in American History: A Reference Guide with Primary Documents.* Westport, Conn.: Greenwood Press, 2006.

Markowitz, Gerald, and David Rosner. *Deceit and Denial: The Deadly Politics of Industrial Pollution.* Berkeley: University of California Press, 2002.

Snyder, Lynne Page. "Revisiting Donora, Pennsylvania's 1948 Air Pollution Disaster." In *Devastation and Renewal: An Environmental History of Pittsburgh and Its Region,* edited by Joel A. Tarr. Pittsburgh: University of Pittsburgh Press, 2003.

# Dust Bowl

CATEGORIES: Disasters; land and land use; resources and resource management

THE EVENT: Environmental disaster marked by huge dust storms in the southern region of the Great Plains of the United States

DATE: 1930's

SIGNIFICANCE: The Dust Bowl revealed the damage that mechanized agriculture could cause if not accompanied by a program of soil management.

Droughts periodically occur in the Great Plains of the United States. During such periods, winds pick up loose soil and create dust storms, especially during the spring months. Settlers reported numerous examples of this natural phenomenon during the nineteenth century. During the twentieth century, new agricultural practices and overgrazing by cattle speeded soil erosion in the region. Tractors and other machines allowed farmers to plow larger areas for planting wheat. In the process, they destroyed the natural grasses, the root systems of which had stabilized the soil. Because the wheat replaced the grasses, most farmers remained unaware that they were contributing to a coming catastrophe.

In 1931 a severe drought struck the Great Plains; it centered on the Texas and Oklahoma panhandles, northeastern New Mexico, eastern Colorado, and southwestern Kansas. The wheat crop withered in the fields, and its root systems were no longer able to support the soil. As the drought continued, soil particles that normally clustered together separated into a fine dust. When the winds blew in early 1932, they lifted the dust into the air, marking the beginning of the environmental disaster that a newspaper reporter later dubbed the Dust Bowl.

Although their number and severity increased, dust storms remained an issue of local and regional concern for the first two years. However, as the drought continued into 1934, the storms grew so immense that they caused damage in areas far from the plains. A storm that emanated from Montana and Wyoming in May, 1934, deposited an estimated twelve million tons of dust on Chicago, Illinois. Ships that were some 480 kilometers (300 miles) offshore in the Atlantic Ocean reported that dust from the same storm landed on their decks. Incidents such as these provoked national concern over the growing crisis on the plains.

Scientists identified two types of dust storms: those caused by winds from the southwest and those resulting from air masses moving from the north. While no less damaging, the more frequent southwest storms tended to be milder than the terrifying northern storms, which came to be known as "black blizzards." Huge walls of dust, sometimes more than 1.6 kilometers (1 mile) high, rolled across the plains at 100 kilometers per hour (62 miles per hour) or faster, driving frightened birds before them. The sun would disappear, it would become as dark as night, and frightened people would huddle in their homes, their windows often taped shut. On occasion, people stranded outside during these severe storms suffocated. Some black blizzards lasted less than one hour; others reportedly continued for longer than three days.

Most historians argue that the Dust Bowl was one of the worst ecological disasters in the United States, one that could have been mitigated had farmers practiced soil conservation in the years before drought struck. Instead, farms were ruined, causing some 3.5 million people to abandon the land. Many of them moved into small towns on the plains, while others journeyed to California in search of opportunity. Cattle and wildlife choked to death. Human respiratory illnesses increased markedly during the Dust Bowl era, and a number of people died from an ailment known as dust pneumonia. Anecdotal evidence indicates that many people grew depressed as the dust storms continued year after year.

The mid-1930's marked the peak of the Dust Bowl, with seventy-two storms that reduced visibility to less than 1.6 kilometers (1 mile) reported in 1937. The return of the rain in the late 1930's eased the crisis, and by 1941 the disaster was over. However, by that time ecologists and farmers had begun to undertake soil conservation measures in response to the crisis. The U.S. government provided expertise and financial support for many of these efforts. Farmers practiced listing, a plowing process that makes deep furrows to capture the soil and prevent it from blowing. Alternating strips of planted wheat with dense, drought-resistant feed crops such as sorghum slowed erosion by blocking wind and retaining moisture, which prevented the soil from separating into dust. On lands not farmed, natural grasses were planted to prevent erosion. The government also sponsored the Shelterbelt Project, a program that used rows of trees to form windbreaks. Millions of trees were planted throughout the Great Plains, with more than 4,828 kilometers (3,000 miles) of shelterbelts created in Kansas alone.

Despite the experiences of the 1930's, once the drought ended many farmers returned to the farming practices that had damaged their fields. Soil conservation experts worried that the region would suffer a return of Dust Bowl conditions when the rains stopped. Their predictions came to pass in 1952, when another drought led to a series of dust storms, including several storms with wind gusts clocked at 129 kilometers (80 miles) per hour. That drought ended in 1957, but in accord with a twenty-year cycle, the region again faced a shortage of rainfall in the early 1970's. At that time some analysts confidently predicted that dust storms such as those seen in the 1930's were a thing of the past. They claimed that irrigation with aquifer water from deep wells would pre-

*A farmer's son sits on a sand dune near his home in Liberal, Kansas, during the 1930's.* (Library of Congress)

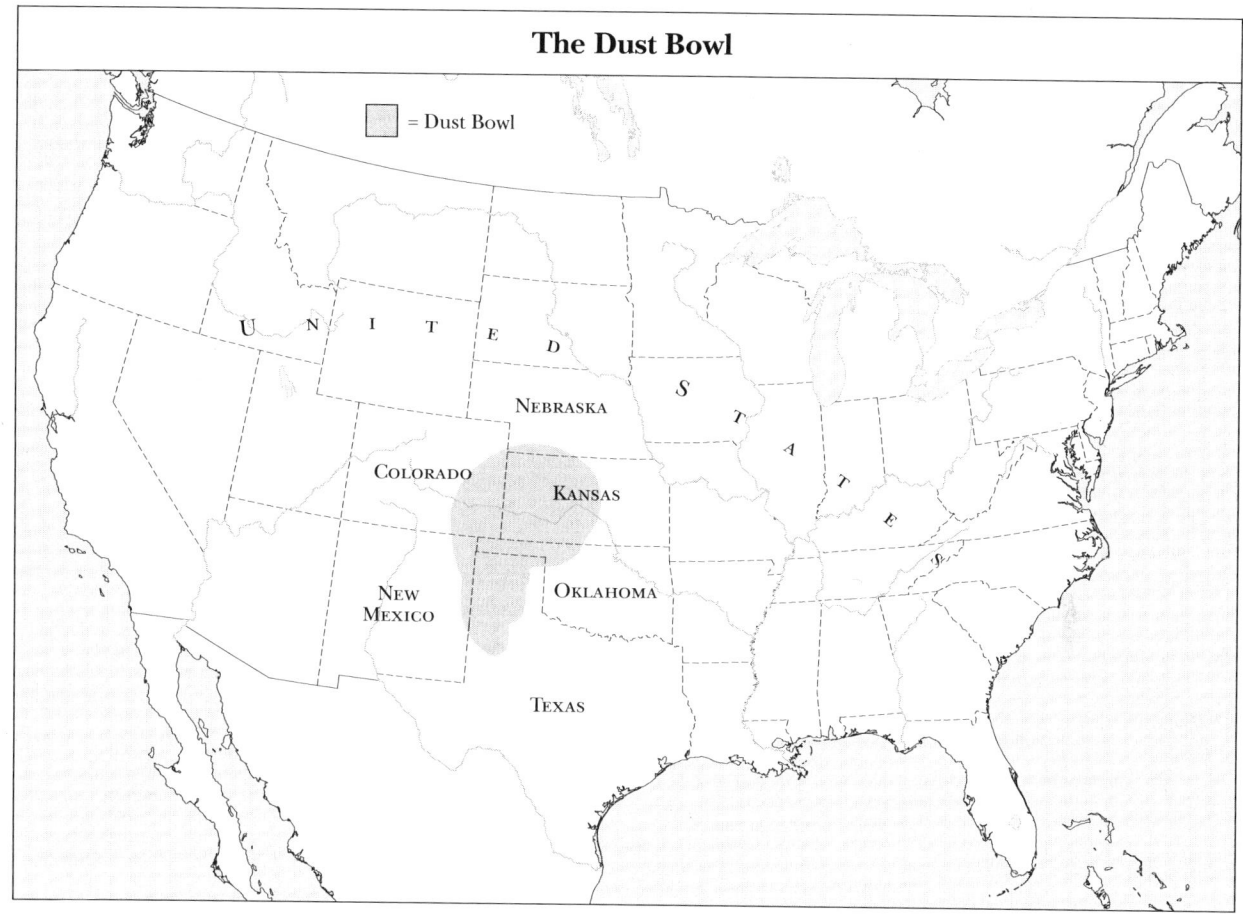

**The Dust Bowl**

☐ = Dust Bowl

UNITED STATES

NEBRASKA

COLORADO

KANSAS

NEW MEXICO

OKLAHOMA

TEXAS

vent soil erosion. However, shrewd observers pointed out that the fate of the region was now tied to a resource, aquifer water, that would become increasingly precious in the coming years. The possibility that the Great Plains could again witness a devastating ecological catastrophe like that of the 1930's remains.

*Thomas Clarkin*

FURTHER READING

Bonnifield, Paul. *The Dust Bowl: Men, Dirt, and Depression.* Albuquerque: University of New Mexico Press, 1979.

Cunfer, Geoff. *On the Great Plains: Agriculture and Environment.* College Station: Texas A&M University Press, 2005.

Hurt, R. Douglas. *The Dust Bowl: An Agricultural and Social History.* Chicago: Burnham, 1981.

Lookingbill, Brad D. *Dust Bowl, USA: Depression America and the Ecological Imagination, 1929-1941.* Athens: Ohio University Press, 2001.

Worster, Donald. *Dust Bowl: The Southern Plains in the 1930s.* 25th anniversary ed. New York: Oxford University Press, 2004.

# El Niño and La Niña

CATEGORY: Weather and climate

DEFINITION: El Niño is a quasi-periodic abnormal warming of surface waters in the central and eastern tropical Pacific Ocean; La Niña is a quasi-periodic abnormal cooling of surface waters in the central and eastern tropical Pacific Ocean

SIGNIFICANCE: The El Niño-Southern Oscillation climate pattern has implications for weather around the world. El Niño and La Niña events affect temperatures and precipitation patterns in many different regions and are associated with extreme weather such as heavy rainfall, floods, and droughts.

Ecuadoran and Peruvian fishermen gave the name El Niño to a warm, southward-flowing ocean current that would occur off the west coasts of their countries every year around Christmastime (*El Niño* is Spanish for "the little boy" or, more specifically, "the Christ child"). Originally used to describe a brief, localized, annual phenomenon, the term later became associated with unusually strong ocean warming in this area that occurred every few years, disrupting local fish and bird populations. "Anti-El Niño" ocean cooling events were called either La Niña (the little girl) or El Viejo (the old man).

The term "El Niño" has come to be associated with the large warm-water anomalies covering extensive portions of the tropical Pacific Ocean off the coast of Latin America that persist for many months, and with the related weather effects noted around the globe. It has been known for many years that when warm water appears off the coast of Peru, atmospheric pressure drops over the eastern Pacific and rises over Australia and the Indian Ocean. Because of this relationship, major El Niño events are usually associated with other global weather phenomena, including drought in Africa and Australia and the failure of the Indian monsoon. Scientists know the atmospheric component of this global pattern as the Southern Oscillation. The coupled global oceanic-atmospheric system is called El Niño-Southern Oscillation (ENSO). El Niño is sometimes called an ENSO warm event or the warm phase of ENSO; similarly, La Niña may be referred to as an ENSO cold event or the cold phase of ENSO.

Typically, El Niño and La Niña events occur every three to five years, although the interval may vary from two to seven years. El Niño events tend to last nine to twelve months but have been known to last as long as two, three, or even four years. The typical La Niña event lasts one to three years. Both El Niño and La Niña conditions typically develop during the months of March through June, reach peak intensity during the months of December through April, and diminish during the months of May through July.

CAUSES AND PREDICTION

When an El Niño condition develops in the eastern Pacific, the sea surface temperature and rainfall in the eastern tropical Pacific are at their seasonal peaks. Major El Niño occurrences are closely tied to global weather patterns and the circulation of currents in the Pacific Ocean. Variations in Indian monsoon circulation sometimes precede variations in the Southern Oscillation, indicating that there is a possible feedback mechanism linking these phenomena. The period of the Southern Oscillation is irregular, with a return period of about three to four years, so about two El Niños occur per decade. The amplitude of the Southern Oscillation is highly irregular. If some global atmospheric perturbation contributes to the amplitude of the Southern Oscillation while an El Niño is developing, a major El Niño might be expected to occur. However, if a global perturbation subtracts from the amplitude of the Southern Oscillation, the El Niño might be weak.

The point at which scientists decide that a major El Niño condition is occurring has been historically contentious. A network of buoys has been established to augment satellite monitoring of sea surface temperatures in the Pacific Ocean. This network played a key role in the early detection of the 1997-1998 El Niño. When sea surface temperatures reach 3 to 5 degrees Celsius (5.4 to 9.0 degrees Fahrenheit) above normal for the season in the eastern equatorial Pacific, scientists can be fairly certain that an El Niño is occurring. The classic El Niño begins off the coast of Peru, slowly propagating westward. Since the 1990's, however, a new type of El Niño has been observed in which the maximum ocean warming occurs instead in the cen-

## Years in Which El Niño and La Niña Were Observed, 1900-2007

| EL NIÑO YEARS<br>(warm water in eastern Pacific) | LA NIÑA YEARS<br>(cold water in eastern Pacific) |
|---|---|
| 1902, 1905, 1911, 1914, 1918, 1923, 1925, 1930, 1932, 1939, 1941, 1951, 1953, 1957, 1965, 1969, 1972, 1976, 1982, 1986, 1991, 1994, 1997, 2002, 2004, 2006, 2009 | 1904, 1908, 1910, 1916, 1924, 1928, 1938, 1950, 1955, 1964, 1970, 1973, 1975, 1988, 1995, 1998, 2007, 2010 |

*Note:* Many El Niños begin in one calendar year and end during the following calendar year. Only the beginning year is listed.

tral-equatorial Pacific. Known variously as a central Pacific El Niño, dateline El Niño, warm-pool El Niño, or El Niño Modoki, this type of event occurred in 1991-1992, 1994-1995, 2002-2003, 2004-2005, and 2009-2010.

During a major El Niño, the normal westerly trade winds subside, and the height of the eastern Pacific sea surface rises. This is coupled with a decline in the height of the western Pacific sea surface, which sometimes causes normally submerged coral reefs in the western Pacific to appear above the ocean surface.

Since historical recording of El Niño events began, long-term variations in their strength have been observed. The 1920's and 1930's experienced only weak El Niño events. In contrast, the El Niños of the 1980's and 1990's were generally moderate to strong. El Niño conditions result from a complex interplay of atmospheric and oceanic forces, and the reasons for the waxing and waning in strength of El Niño events over periods of decades are not understood.

Some scientists have noted that unusual El Niños have followed volcanic eruptions. A major volcanic eruption causes the formation of a stratospheric aerosol layer, which may lead to an increase in solar radiation being reflected back into space. The 1951 El Niño followed the eruption of Mount Lamington in Papua New Guinea; the 1982 El Niño followed the eruptions of El Chichon in Mexico and Galunggung in Indonesia; the 1991 El Niño followed the eruption of Mount Pinatubo in the Philippines. However, while 1951 was a weak El Niño year, both 1982 and 1991 were strong. Although it is tempting to focus on a single parameter such as sea surface temperatures or the period of the Southern Oscillation as a predictor of El Niño, history shows that many factors, both atmospheric and oceanic, contribute to the development of a strong El Niño event.

## ENVIRONMENTAL CONSEQUENCES

When unusually warm water appears off the coast of Peru, the local anchovy fishing industry falters. Sportfishing off Baja California, California, and Oregon, in contrast, enjoys a boom, as marlin and other highly prized fish usually found in more tropical southern waters move north. Other marine ecosystems around the globe may also be affected by El Niño conditions. Factors such as increased sea surface temperature, decreased sea level, and salinity changes related to high rainfall affect the algae that protect coral reefs, causing the coral to bleach and die. This, in

turn, has negative impacts on fish and plant life within reef ecosystems. The 1997-1998 El Niño event was marked by coral reef bleaching in the Indo-Pacific region, the Caribbean, and the Florida Keys. Extensive coral bleaching also occurred around the globe in association with the 2009-2010 El Niño.

During El Niño episodes, the intertropical convergence zone, a band of major tropical convection circling the globe, moves southward. This southward shift in precipitation patterns causes torrential rains in some places that are normally dry and dry conditions in places that are usually wet. In the Galápagos Islands, El Niño brings much higher than normal precipitation in March, April, and May. During major El Niños, Peru and Ecuador experience torrential rains and flooding. In Guayaquil, Ecuador, El Niño was blamed for causing more than 3 meters (9.8 feet) of rain between October, 1982, and January, 1983. During the 1997-1998 El Niño, severe flooding occurred in Ecuador along rivers where rain forests had been cleared to establish shrimp farms. By contrast, the moderate El Niño of 2002-2003 had little effect on the weather of these two countries.

During many major El Niño events, countries bordering the western Pacific and Indian oceans experience droughts. In El Niño years India often receives lower-than-normal rainfall, while Sri Lanka's rainfall tends to be unusually high. During the 1982-1983 El Niño, Indonesia and Australia were stricken by drought. Curiously, the Australian drought associated with the strong 1997-1998 El Niño was not as severe as the one that accompanied the moderate 2002-2003 El Niño. Early in the 1990's southern Africa experienced its worst drought of the twentieth century, probably worsened by the El Niño that began in late 1991. El Niño years in Japan are associated with mild winters, cool summers, and lengthy rainy seasons.

Many diverse ecological, environmental, and economic events throughout the world are often attributed to El Niño occurrences. Sometimes these events may indeed be related to El Niño, but some occurrences can be attributed to other factors. Sometimes conflicting claims are made about the effects of El Niños. Just as there is no consistent relationship between the failure of the Indian monsoon and El Niño years, other claims about El Niño and weather may hold up during a statistically significant number of years but not in all years.

Although the 1997 El Niño was credited with causing the unusually mild winter of 1997-1998 in the eastern United States, the 1976 El Niño was blamed for causing extreme cold in the same region in December, 1976, and January, 1977. The Sonoran Desert of Arizona and California tends to be wet in El Niño years. The Florida drought of 1998 was attributed to El Niño, although the warm-water anomaly off Peru had virtually disappeared by July, when forest fires were plaguing Florida. Because Texas often experiences more precipitation during the growing season after an El Niño, farmers planted crops requiring more moisture than usual in 1998. The summer of 1998 was unusually dry in Texas and Oklahoma, however, leaving many of the affected farmers to conclude that El Niño-based forecasts might be less than reliable. The failure of the Soviet harvest in 1972 was attributed by some to El Niño. In El Niño years, Moscow frequently experiences very cold winters; December, 1997, fit into this pattern.

## La Niña

As the warm-water anomalies in the eastern Pacific fade, scientists know that El Niño is ending. When a large body of colder-than-average water establishes itself off the coast of Peru, along with a strong ridge of high pressure, meteorologists announce that a La Niña event is occurring. A La Niña may follow closely on the heels of an El Niño, as was the case with the 2010-2011 La Niña, or occur after a year or two of neutral conditions. During a La Niña, the Pacific sea surface height in the eastern Pacific is measurably lower than when an El Niño is occurring. The westerly trade winds are also stronger than normal. During a La Niña, the average temperature of the tropical troposphere may be 1 degree Celsius (1.8 degrees Fahrenheit) lower than during an El Niño. La Niñas tend to be much more variable in strength than El Niños.

During periods when the waters of the eastern Pacific have been observed to be anomalously cold, the Pacific Northwest tends to be wetter and cooler than normal, especially during winter. During a La Niña year, winter temperatures in the southern and southeastern United States are often warmer than normal. In the United States, some link La Niña years to very hot summers; the summer of 1988, which was very hot and dry, is the prototype summer for this weather phenomenon. Tropical cyclones are more common on the northern Australian coast during La Niña events. Widespread flooding in eastern Australia tends to occur early in the calendar year (late summer) during La Niñas. La Niña conditions also tend to be accom-

panied by more damaging Atlantic hurricanes, while a reduction in hurricane activity is typically associated with El Niño.

*Anita Baker-Blocker*
*Updated by Karen N. Kähler*

## Further Reading

Changnon, Stanley Alcide, ed. *El Niño, 1997-1998: The Climate Event of the Century.* New York: Oxford University Press, 2000.

Clarke, Allan J. *An Introduction to the Dynamics of El Niño and the Southern Oscillation.* San Diego, Calif.: Academic Press, 2008.

D'Aleo, Joseph S., and Pamela G. Grube. *The Oryx Resource Guide to El Niño and La Niña.* Westport, Conn: Oryx Press, 2002.

Glantz, Michael H. *Currents of Change: Impacts of El Niño and La Niña on Climate and Society.* 2d ed. New York: Cambridge University Press, 2001.

_____. *La Niña and Its Impacts: Facts and Speculation.* Tokyo: United Nations University Press, 2002.

Philander, S. George. *Our Affair with El Niño: How We Transformed an Enchanting Peruvian Current into a Global Climate Hazard.* Princeton, N.J.: Princeton University Press, 2006.

Rosenzweig, Cynthia, and Daniel Hillel. *Climate Variability and the Global Harvest: Impacts of El Niño and Other Oscillations on Agroecosystems.* New York: Oxford University Press, 2008.

# Europe

CATEGORIES: Ecology and ecosystems; resources and resource management; atmosphere and air pollution; water and water pollution

SIGNIFICANCE: Because they are heavily populated and highly industrialized, the European nations have significant impacts on the global environment. The decisions made by Europeans regarding the management and use of natural resources have the potential to affect the world's environment in either positive or negative ways.

Under the leadership of the European Union, Europe is intensely involved in efforts to reduce the environmentally harmful effects on the earth that result from human activity in modern industrialized societies. The European countries address these issues

## Habitats of Selected Vertebrates of Europe

not only from a European perspective but from a global perspective as well. European policy makers are highly aware of their role in the global community and of the impacts of their decisions and activities on the global environment. For example, as the leaders of industrialized nations, they have accepted a serious commitment to reduce greenhouse gas emissions, which have been determined to be a significant cause of global warming.

Because of the diversity of terrain and the differences in natural resources from one country to another as well as varying country locations, some having major coastal areas and others being totally landlocked, environment-related priorities vary among European nations. All share concerns regarding several major environmental issues, however; these include greenhouse gas emissions and climate change, changes in land use and deforestation, loss of biological diversity, rising sea levels, air pollution, and water pollution (especially eutrophication, or the depletion of oxygen in water).

## GREENHOUSE GAS EMISSIONS AND CLIMATE CHANGE

Global warming is a major concern among Europeans. All of the European countries have signed and ratified the Kyoto Protocol, which was set forth by the United Nations Framework Convention on Climate Change. The objective of the agreement is to combat global warming by reducing the amount of greenhouse gases emitted into the air by setting binding target reductions in emissions for each country. European nations have concentrated on the reduction of fossil-fuel use as the primary means of meeting their Kyoto Protocol commitments.

Because coal-fired power plants are the most significant emitters of carbon dioxide ($CO_2$), the primary greenhouse gas, the reduction and even elimination of the use of coal as a fuel is a major goal in Europe; this trend is supported by environmental groups, including Green political parties. The coal industry, particularly in England, which has had a significant history of coal mining, has attempted to avoid being eliminated completely by developing ways to make coal a cleaner-burning fuel. Two methods that have been suggested are the underground gasification of coal and the trapping and storing of $CO_2$ emissions underground.

Hydropower has become an important fuel source for power plants in many European nations, especially in countries, such as Norway, that have abundant rivers and lakes. Europe has also investigated the feasibility of using wind and solar power. Another major way in which Europe is attempting to reduce greenhouse gas emissions is by reducing the use of fossil fuels (oil in the form of gasoline and diesel fuel) in the transportation industry. An increase in the number of transport vehicles has resulted in an increase in fuel consumption, which has the potential to nullify previous reductions made in emissions from power plants and other sources. Biofuels have been proposed as alternatives to fossil fuels.

The European Commission (the executive body of the European Union) has made two important proposals regarding the use of renewable resources for the production of energy. The commission has suggested that a target be set requiring that 20 percent of energy used in Europe be produced using renewable resources by 2020. In addition, the commission has proposed that as much as 10 percent of road transport fuels should be biofuels by 2020. Debates continue about the attainability of these goals, and some observers have expressed concerns about the economic effect of increased biofuel production on food-crop production as well as about the environmental impacts of changes in land usage. These concerns do not necessarily make the use of biofuels infeasible, however; producing biofuels from crop and forest residues and wastes would not have adverse effects on the environment.

In regard to the production of biofuels, European nations have considered the effects of such production on ecosystems in other parts of the world. They are aware that the reduction of food-crop production in Europe would necessitate increased production elsewhere to meet the global need, and that this could affect land use, possibly resulting in the destruction of wetlands and rain forests.

Another environmental concern of the nations of Europe is the projected rise in sea levels as a result of global climate change. Rises in sea levels cause coastal deterioration and could result in significant loss of landmass, posing a potential economic threat in the areas of tourism and recreation. The flooding that would result from rising sea levels is also a major concern, particularly to the Netherlands, large portions of which lie near or below sea level. In addition to implementing projects for raising dikes and securing seawalls, the Netherlands has considered dumping millions of tons of sand into the sea to extend its coastline.

## CHANGES IN LAND USE AND DEFORESTATION

The relationships of land usage to biodiversity and to air and water pollution are major environmental concerns in Europe. Any activities that alter or destroy ecosystems result in the loss or serious reduction in numbers of the plant and animal species that depend on those ecosystems for habitat. It has been argued that the conversion of increasing areas of arable land to the planting of crops to meet the needs of both food-crop and biofuel-crop production could pose serious threats to biodiversity throughout Europe. Increasing the land used for crop production would have serious impacts on forests and grasslands. Also, the destruction of these types of land areas would remove substantial quantities of the major natural means of removing $CO_2$ from the air—that is, plants and trees.

The environmental impacts of farming constitute a significant concern among Europeans. Almost one-half of the land area within the European Union is used for agriculture; thus farming methods and animal husbandry have major effects on the European environment. With the increased global demand for food production in response to population growth, farming methods have become more intensive in Europe just as they have worldwide. This has resulted in increased use of fertilizers and pesticides as well as increased land usage for crop production, with serious impacts on both air quality and water quality.

Runoff from fertilized fields has increased the amounts of nutrients in waterways and in surrounding seas, causing harmful effects on both freshwater species and sea life. Residues from pesticides have increased the amounts of toxins in both air and water, destroying aquatic life and causing severe damage to forests. Because of their location, the forests of Switzerland in particular have suffered severe damage from air pollution, with almost 35 percent of the trees seriously affected. Through its system of agricultural programs known as the Common Agriculture Policy, the European Union has implemented reforms to ameliorate these situations throughout Europe. In addition, the agricultural industry has worked to combat the environmentally adverse effects of farming.

## Loss of Biological Diversity

Farming, industrialization, construction of roads and recreational areas, and urban sprawl have all contributed to loss of habitat and its accompanying loss of biodiversity throughout Europe. Although the nations of Europe did not meet a 2010 target for halting loss of biodiversity, significant improvement has been seen. More and more land in Europe has been protected as habitat through both European Union programs and individual country provisions. The Natura 2000 network provides protection from exploitation to 17 percent of the land included in the European Union.

Thirty-nine European countries have programs that ensure protection of habitat, but European biodiversity is still very susceptible to loss, with from 40 percent to 85 percent of habitat and 40 percent to 70 percent of plant and animal species estimated to be at risk. Owing to the ease with which they can be converted to cropland, grasslands and wetlands are at high risk of being lost throughout Europe. Increased use of hydropower also poses a potential threat to biodiversity. Dams and reservoirs destroy both habitats and wildlife populations in the areas they flood; fish populations are threatened by the changing of the flow of rivers caused by dams. Some European countries (including Germany, Norway, and Sweden) have enacted measures to address and alleviate these threats, such as through the installation of fish ladders at dams and the use of mini-hydropower stations.

The loss of marine biodiversity is another environmental concern among the European countries. The three major causes of declines in fish and shellfish populations are global warming, pollution, and overexploitation by the fisheries industry. Global warming has caused certain species of fish to move to cooler waters in more northern regions and has also interfered with spawning owing to unfavorable habitat conditions. Farming and other practices of modern industrialized societies have contributed to declines in populations as well, as agricultural runoff containing pesticides, industrial chemicals, oil spills, and other toxins have caused marked reductions in marine life. Nutrients contained in runoff water from artificially fertilized fields also create unfavorable conditions for fish. In deep-sea waters, these nutrients produce the highly toxic substance hydrogen sulfide.

Overexploitation of the seas in the form of the excessive taking of fish by commercial fisheries is a very significant cause of great reductions in both numbers and species of fish. The European Union addresses this problem through the Common Fisheries Policy, which limits the numbers of fish of particular species that may be taken based on the results of scientific studies. Critics have argued, however, that the importance of the fisheries industry to the economies of many European nations has led to the setting of these limits at levels that tend to be higher than are actually environmentally sound. This situation, coupled with the numbers of fish taken illegally, continues to present a problem in Europe's reestablishment of aquatic populations.

*Shawncey Webb*

FURTHER READING

Blennow, Kristina, ed. *Sustainable Forestry in Southern Sweden: The Sufor Research Project.* New York: Haworth Press, 2006.

Deublein, Dieter, and Angelika Steinhauser. *Biogas from Waste and Renewable Resources: An Introduction.* Weinheim, Germany: Wiley-VCH, 2008.

Glover, Leigh. *Postmodern Climate Change.* New York: Routledge, 2006.

Hasenauer, Hubert, ed. *Sustainable Forest Management: Growth Models for Europe.* Berlin: Springer-Verlag, 2006.

Maslin, Mark. *Global Warming: A Very Short Introduction.* 2d ed. New York: Oxford University Press, 2009.

Thorsheim, Peter. *Inventing Pollution: Coal, Smoke, and Culture in Britain Since 1800.* Athens: Ohio University Press, 2006.

# Federal Insecticide, Fungicide, and Rodenticide Act

CATEGORIES: Treaties, laws, and court cases; human health and the environment

THE LAW: U.S. legislation that established the role of the federal government in regulation of pesticides

DATE: Enacted on June 25, 1947

SIGNIFICANCE: The Federal Insecticide, Fungicide, and Rodenticide Act established firm guidelines for the registration and use of pesticide products in the United States, resulting in a significant reduction in harm to humans and other animals from exposure to pesticides.

The Federal Insecticide, Fungicide, and Rodenticide Act of 1947 (FIFRA) represented the culmination of events in which the food and chemical industries in the United States began a merger of practices. In the late nineteenth century, American consumers began to be alarmed—with some justification—by farmers' practice of applying insecticides and other toxic chemicals to crops haphazardly. The Insecticide Act of 1910 and various so-called pure food and drug laws were early attempts to regulate such chemicals.

During the 1940's, a large number of toxic chemicals were developed as a result of the war effort. With the end of World War II, industries attempted to find other uses for chemicals that were no longer in short supply. Some of the chemicals had potential applications as insecticides or pesticides, and it was logical for the manufacturers to look for uses for them in the food industry. At the time, it was estimated that nearly 200 million bushels of corn and small grain could be added to the market if a population of some 140 million rats could be brought under control.

With FIFRA's passage by Congress in 1947, the federal government began to oversee the labeling and marketing of all pesticides and fungicides entering the U.S. marketplace. The act required a scientific review by members of the Department of Agriculture of any chemicals introduced into the interstate market and also required such chemicals to be packaged with proper labeling of claims and precautions. Revisions to FIFRA during the 1970's strengthened federal oversight of pesticides and fungicides, including requiring the registration of new products with the Environmental Protection Agency.

*Richard Adler*

# Freon

CATEGORY: Atmosphere and air pollution

DEFINITION: A nontoxic, nonflammable refrigerant gas

SIGNIFICANCE: Freon and other chlorofluorocarbons served a number of purposes in many industries, but it was found that they were harmful to the earth's ozone layer, and nations around the world united in banning their use.

Freon is the Du Pont Corporation's trade name for a compound used as a refrigerant. Freon, which was introduced in 1930, is an example of a class of gases known as chlorofluorocarbons (CFCs), carbon compounds that contain fluorine and chlorine. They are derivatives of simple alkanes—such as methane, ethane, propane, and butane—through direct or selective ultraviolet halogenation using chlorine or fluorine gas.

Freon found extensive uses in industry. CFCs have served as dispersing gases in aerosol cans, in the preparation of foamed plastics, and, primarily, as refrigerants. Their manufacture, together with that of the closely related halons, rose in the mid-1970's and peaked in 1986 with the production of almost 1.25

million tons. At that time these compounds were universally used in aerosol products ranging from insecticides to shaving foams and hair sprays, as well as in the insulation of buildings and as cleaning solutions for circuit boards and other electronic parts. One of them, bromotrifluoromethane (Freon 13B1), was used as a fire extinguisher in situations where the use of water had to be avoided, such as electrical fires.

Common members of this family of chemicals include trichlorofluoromethane (Freon or CFC 11), dichlorodifluoromethane (CFC 12), and 1,2-dichloro 1,1,2,2-tetrafluoroethane (CFC 114). They are all either gases or low-boiling liquids at room temperature and are virtually insoluble in water. They are generally dense, easily liquefied, not flammable, thermally stable, virtually odorless, and inexpensive to manufacture. They do not undergo decomposition via the ordinary chemical reactions that take place in the troposphere. As a result, they were seen as ideal for use as propellants in aerosol cans of deodorants, hair sprays, and various commercially available food products. Their relative inertness toward other chemicals allows them also to persist in the atmosphere, causing environmental problems.

Because they are water-insoluble, rain cannot dissolve them and wash them down to the ground. As a result, they drift upward into the stratosphere and the ozone layer, which they reach after approximately seven to ten days. They may stay in the stratosphere for several decades, absorb the sun's ultraviolet light, and yield free radicals, which appear to undergo chemical reactions that lead to the depletion of the ozone layer. Although ozone is toxic to human lungs, its presence in the stratosphere is critical in protecting the earth from the harmful ultraviolet part of the electromagnetic radiation associated with sunlight. If the ozone layer gets thin, exposure to ultraviolet radiation will exponentially increase the cases of skin cancer and other diseases while at the same time destroying crops and other plants.

CFCs have been found to escape into the atmosphere from old refrigerators and air-conditioning units. Most industrialized countries have banned their use and have replaced them with methylene chloride or nonhalogenated hydrocarbons, such as isobutane. The flammability of those hydrocarbons and the suspected carcinogenicity of methylene chloride have created an incentive for the development of CFC substitutes. In 1970 the U.S. Congress passed legislation aimed at curbing the sources of air pollution by setting standards for air quality. In 1987 more than twenty nations signed the Montreal Protocol, an agreement to downscale production of CFCs, with the intent of eventually eradicating their use. By 2009, all members of the United Nations had ratified the protocol, and it was expected that if all nations comply with their obligations under ongoing revisions of the protocol, the ozone layer will recover by 2050.

*Soraya Ghayourmanesh*

FURTHER READING

Dauvergne, Peter. "Refrigerating the Ozone Layer." In *The Shadows of Consumption: Consequences for the Global Environment.* Cambridge, Mass.: MIT Press, 2008.

Joesten, Melvin D., John L. Hogg, and Mary E. Castellion. "Chlorofluorocarbons and the Ozone Layer." In *The World of Chemistry: Essentials.* 4th ed. Belmont, Calif.: Thomson Brooks/Cole, 2007.

Parson, Edward A. *Protecting the Ozone Layer: Science and Strategy.* New York: Oxford University Press, 2003.

# Global warming

CATEGORY: Weather and climate

DEFINITION: Increase in the average surface and ocean temperature of the earth since 1850 and the projected persistence of the trend

SIGNIFICANCE: The findings of scientists concerning the causes of global warming are extremely important in that they provide guidance for policy makers. Harmful consequences may result if the anthropogenic, catastrophic theory of global warming is correct and policy makers do not take the political and economic actions necessary to address the problem; conversely, harmful consequences may result if the theory is wrong but major political and economic decisions are made in the belief that it is correct.

According to the Intergovernmental Panel on Climate Change (IPCC), the overall global temperature during the twentieth century increased by a little less than 1 degree Celsius (1.8 degrees Fahrenheit). This involved an increase of about 0.5 degree Celsius (0.9 degree Fahrenheit) from 1910 to 1945 and a similar increase from about 1975 to 2000 or so (actually peaking in 1998), with a slight decrease in the intervening years. (The figures are approximate because of uncertain data and yearly fluctuations, occasionally as large as 0.25 degree Celsius up or down, and the complexity of adjusting the raw temperature data.) Explaining these increases and projecting future trends and their consequences are the key issues addressed by scientists who examine global warming.

## CLIMATE CYCLES AND HUMAN CAUSES

Two basic theories have been posited regarding the source of the warming, both of which could easily be partially correct. Some see the warming as basically natural (as many scientists agree is probably the case for the pre-1945 warming, which predates the large increase in atmospheric carbon dioxide). In fact, short-term natural causes have been documented, such as volcanic eruptions (the Mount Pinatubo eruption in 1991 was followed by a strong temperature down-spike in 1992) and the El Niño/La Niña weather cycle (the very strong El Niño of 1998 resulted in a large temperature up-spike). In addition, some long-term fluctuations, such as the Pacific Decadal Oscillation and the Atlantic Multidecadal Oscillation, affect global as well as local temperatures. In addition, solar energy is not constant; there are slight cyclical variations that correlate with sunspot activity. These do not seem to be sufficient to explain the post-1975 warming; Patrick J. Michaels and Robert C. Balling, Jr., in their 2009 book *Climate of Extremes: Global Warming Science They Don't Want You to Know*, estimate that natural causes explain only 25 percent of the post-1975 warming (compared to 75 percent of the earlier warming). Some scientists, however, think natural answers can be found for the rest.

These scientists think the current warming is natural and cyclic, a Modern Warm Period to follow the Medieval Warm Period and Little Ice Age (which ended about 1850). A wide array of historical temperature proxies show that there is a roughly 1,500-year cycle, with shorter heating and warming subcycles. The proxies include ice cores dating back hundreds of thousands of years, six thousand boreholes (from all continents), seabed and lake-bed sediment cores, tree rings and tree lines, cave stalagmite cores, peat bogs, and records. These do show occasional remarkable shifts, global temperature increasing nearly 1 degree Celsius for about a decade at the end of the Younger Dryas (11,500 years ago) for reasons still unknown (there was an increase in greenhouse gases after the rise).

Environmental scientist S. Fred Singer has estimated that the Medieval Warm Period exceeded the current warming (so far). Others dispute this. Clima-

---

## The IPCC's 2007 Assessment of Climate Change

*On its Web site, the Intergovernmental Panel on Climate Change summarized the following findings of its 2007 assessment report:*

- Warming of the climate system is unequivocal.

- Most of (50% of) the observed increase in globally averaged temperatures since the mid-20th century is very likely (confidence level 90%) due to the observed increase in anthropogenic (human) greenhouse gas concentrations.

- Hotter temperatures and rises in sea level "would continue for centuries" even if greenhouse gas levels are stabilized, although the likely amount of temperature and sea level rise varies greatly depending on the fossil intensity of human activity during the next century.

- The probability that this is caused by natural climatic processes alone is less than 5%.

- World temperatures could rise by between 1.1 and 6.4° Celsius (2.0 and 11.5° Fahrenheit) during the twenty-first century. . . .

- Sea levels will probably rise by 18 to 59 centimeters (7.08 to 23.22 inches).

- There is a confidence level 90% that there will be more frequent warm spells, heat waves and heavy rainfall.

- There is a confidence level 66% that there will be an increase in droughts, tropical cyclones and extreme high tides.

- Both past and future anthropogenic carbon dioxide emissions will continue to contribute to warming and sea level rise for more than a millennium.

- Global atmospheric concentrations of carbon dioxide, methane, and nitrous oxide have increased markedly as a result of human activities since 1750 and now far exceed pre-industrial values over the last 650,000 years.

tologist Michael E. Mann has argued that global temperatures changed only slightly during these periods, far less than in the twentieth century.

Among the many possible anthropogenic, or human-caused, influences on climate change is land use. The effects of land use are important, but they are for the most part local; cropland is warmer than forest, and urban areas are much warmer than cropland (the urban heat island effect). Overgrazing can lead to desertification, which makes the land warmer.

The strongest anthropogenic effect comes from the production of greenhouse gases such as carbon dioxide. Water vapor is an extremely important greenhouse gas, and methane (much of it from rice paddies and livestock raising) is far more powerful than carbon dioxide, but a trend toward increasing atmospheric methane halted during the mid-1990's. Carbon dioxide is the greenhouse gas that is the cause of greatest concern. Since the beginning of the twentieth century, the amount of carbon dioxide in the atmosphere has increased from about 290 parts per million to almost 390. The increase in the greenhouse effect is far smaller than the increase in greenhouse gases, however, particularly in areas of high humidity, owing to the atmospheric equivalent of the law of diminishing returns. Greenhouse gas warming is strongest at night and therefore in winter, and in upper latitudes; in the atmosphere it leads to a warmer troposphere and cooler stratosphere.

## CLIMATE MODELS

Computer climate models can be used for historical research as well as future projections. Extremely complex general circulation models provide detailed information on how natural warming or greenhouse gas warming is likely to affect climate all over the planet and into the atmosphere, and the models' projections can be tested against observational data. Such testing is as necessary for the findings produced by computer models as for the findings produced by any other scientific experiments; results must be shown to be replicable by others, and the data must be freely available for examination by others. One problem with the testing of data from climate models is that observational data are often too recent (satellite tracking of hurricanes began in 1970, for example, and satellite measurement of Arctic sea ice in 1979) to allow scientists to determine reliably whether changes represent coincidental long-term oscillations or result from the current warming trend.

In the early twenty-first century, most climate models project a linear global surface temperature increase from 2 to 3 degrees Celsius (3.6 to 5.4 degrees Fahrenheit) per century (occasionally much more owing to positive feedback effects, such as increased evaporation leading to increased humidity). Early models exaggerated the warming and could not match the previous history. Later models that added in sulfate aerosols were more accurate, but they failed to predict the absence of net warming since 1998 or the relative lack of warming in the Southern Hemisphere; the problem was that the models used one unknown to check another. Some of the early error may have resulted from negative feedbacks, such as clouds (low-level cumulus and stratus clouds reflect solar light, cooling the planet), which the models often ignored. Climate models predict different specific results from natural and greenhouse gas warming, and many observations (most notably the overall cooling of Antarctica) tend to support the latter, though not entirely.

Michael E. Mann and Lee R. Kump have praised three projections made by James E. Hansen, director of the National Aeronautics and Space Administration's Goddard Institute for Space Studies, when he presented testimony to the U.S. Congress on climate change in 1988. The most severe scenario (A) predicted an increase of just over 1 degree Celsius in the following thirty years, comparable to the high-end projection of 3 degrees per century, but started to diverge from reality within a few years as too high. The middle scenario (B) projected an increase of less than 1 degree, roughly comparable to an increase of 2 degrees per century, and the low scenario (C) projected an increase of about 0.25 degree in thirty years, probably less than 1 degree per century. Scenarios B and C tracked closely with each other, and with the actual observed data, up to 2005. Reports since then (including the Goddard Institute's December, 2009, estimate) show that scenario C has been the most accurate.

## CONSEQUENCES AND SOLUTIONS

Global warming may have many possible effects. The Medieval Warm Period, though beneficial to European and Arctic agriculture, often led to drought elsewhere (including the drought suspected of having caused the collapse of the Native American Anasazi culture). Similar effects can be seen in the twenty-first century; the decline in the snowpack on Africa's Mount Kilimanjaro is apparently more a result of increased local aridity than of global warming.

Warming also leads to a sea-level rise of 1 to 2 centimeters (0.4 to 0.8 inch) per decade, which could increase if the vast Greenland and Antarctic ice packs melt significantly (most models predict more snow in the interiors and more meltwater on the edges of these ice packs, and observations confirm this), which could also seriously alter key ocean currents. Also with warming, warm-weather crops can be grown further north and warm-weather habitats invade cold-weather habitats. Some scientists fear that global warming will lead to more frequent or more severe extreme weather events (particularly tropical cyclones), but there has been no observational evidence of such a trend (for example, North Atlantic hurricanes declined after the severe 2005 season).

Suggested approaches to addressing global warming include both adapting to the heat (and the effects of the heat) when it occurs (and meanwhile devoting resources to solving other problems) and trying to reduce the increase in warming. The latter can have no effect on the natural component of temperature rise and will be unnecessary if the increase is small.

Some proposed solutions aimed at reducing the greenhouse gas emissions linked with global warming are questionable. In particular, the substitution of ethanol for fossil fuels has drawbacks: Growing the crops needed to produce ethanol can in some cases decrease food production or increase cropland at the expense of forestland. Given that forests help to remove carbon dioxide from the atmosphere, the net result may be an increase in greenhouse gases.

Among the most useful and affordable ways to reduce greenhouse gases may be to increase the numbers of hybrid and electric vehicles in relation to gasoline-fueled vehicles, to improve energy conservation by individuals and industries, and to reduce reliance on the burning of fossil fuels for electricity by developing alternative sources of power. Because stronger proposed changes would involve serious economic dislocations, calls for such changes generally include long time lines for achievement. Per-capita carbon dioxide emissions have declined slightly from a 1979 peak; if the trend continues, they will level off with global population around 2050.

*Timothy Lane*

FURTHER READING

Alley, Richard B. *The Two-Mile Time Machine: Ice Cores, Abrupt Climate Change, and Our Future.* Princeton, N.J.: Princeton University Press, 2000.

Dutch, Steven I., ed. *Encyclopedia of Global Warming.* 3 vols. Pasadena, Calif.: Salem Press, 2010.

Fagan, Brian. *The Great Warming: Climate Change and the Rise and Fall of Civilizations.* New York: Bloomsbury, 2008.

Houghton, John Theodore. *Global Warming: The Complete Briefing.* 4th ed. New York: Cambridge University Press, 2010.

Mann, Michael E., and Lee R. Kump. *Dire Predictions: Understanding Global Warming.* New York: DK, 2008.

Michaels, Patrick J., and Robert C. Balling, Jr. *Climate of Extremes: Global Warming Science They Don't Want You to Know.* Washington, D.C.: Cato Institute, 2009.

Singer, S. Fred, and Dennis T. Avery. *Unstoppable Global Warming: Every 1,500 Years.* Updated ed. Lanham, Md.: Rowman & Littlefield, 2008.

Weart, Spencer W. *The Discovery of Global Warming.* Rev. ed. Cambridge, Mass.: Harvard University Press, 2008.

# Green buildings

CATEGORIES: Atmosphere and air pollution; urban environments; pollutants and toxins; resources and resource management

DEFINITION: Structures designed and constructed to increase resource efficiency and reduce negative impacts on human health and the environment

SIGNIFICANCE: Residential and commercial buildings generate more than 30 percent of the world's emissions of carbon dioxide, a greenhouse gas that has been linked to global warming. Green buildings reduce carbon emissions substantially and provide significant environmental benefits by reducing solid waste, efficiently using energy and other resources, reducing air and water pollution, and conserving natural resources.

Although energy efficiency and sustainability were not major concerns at the time, the concept of green buildings originated during the mid-nineteenth century. The Galleria Vittorio Emanuele II in Milan, Italy, designed in 1861, and the Crystal Palace in London, England, built in 1851, both used underground air cooling and roof ventilators to reduce their negative impacts on the environment. New York City's Flatiron Building, constructed in 1902, used deep-set windows to control the interior temperature.

From the 1930's to the 1960's technological advances such as the inventions of reflective glass, structural steel, and air-conditioning resulted in the proliferation of high-rise buildings that consumed huge amounts of cheap fossil fuels. During the 1960's, however, environmental consciousness grew, and visionaries began defining green building. During this period scientist James Lovelock formulated the Gaia hypothesis, a holistic concept of the earth as a single, complex organism. In 1969 landscape architect Ian L. McHarg published *Design with Nature*, which helped define green architecture.

## BEGINNINGS OF THE MOVEMENT

On the first Earth Day, in April, 1970, millions of Americans showed their concern about the environment. The 1973 and 1979 oil crises demonstrated the need for the nation to seek energy from diversified sources and become less dependent on fossil fuels. The U.S. government and many corporations began investing in research into methods of energy conservation and alternative energy sources.

During the 1980's architect Malcolm Wells designed green underground and earth-sheltered buildings. In 1982, physicist Amory Lovins and his wife, environmentalist Hunter Lovins, emphasized the basic green principle of using regional resources in founding their Rocky Mountain Institute, a nonprofit resource policy center that promotes resource efficiency and global security. Beginning during the mid-1980's, popular environmental organizations—such as the Sierra Club, Greenpeace, the Nature Conservancy, and Friends of the Earth—became increasingly active. Growing awareness of the problem of sick building syndrome raised concerns regarding the indoor environments of some workplaces. In 1984 architect William McDonough designed a headquarters building for the Environmental Defense Fund in New York City using a high-performance building approach (the building was completed in 1985). During the late 1980's Pliny Fisk III designed Blueprint Farm—a green agricultural community—in Laredo, Texas, using recycled materials, wind power, and photovoltaic panels.

## MILESTONES DURING THE 1990'S

In 1992, the first local green building program began in Austin, Texas, and the U.S. Environmental Protection Agency (EPA) launched the Energy Star pro-gram, a voluntary energy-efficiency labeling program for consumer products. By 2009 Energy Star labels were appearing on more than sixty product categories and Energy Star ratings had become the standard for major appliances, homes, commercial buildings, and heating systems. By November, 2009, one million Energy Star-qualified homes had been built throughout the United States, resulting in an estimated reduction of greenhouse gas emissions equal to the emissions released by 370,000 gasoline-fueled motor vehicles. Many other countries adopted the Energy Star idea, including Japan, Taiwan, China, New Zealand, South Africa, and the nations of the European Union.

In 1993 Bill Clinton's presidential administration began the successful "Greening of the White House" initiative, and the nonprofit U.S. Green Building Council (USGBC) was created to promote the construction of environmentally responsible, healthy, and profitable buildings. USGBC is a national, voluntary, consensus coalition with members from all sectors of the building industry. In 1995, USGBC began developing its green building certification program, known as LEED (for Leadership in Energy and Environmental Design), which became available for public use in 2000. This voluntary system provides third-party certification that certain standards have been met in the construction of high-performance, sustainable buildings, with an emphasis on reducing carbon dioxide emissions and increasing energy efficiency. LEED certification covers a wide range of existing and new commercial and residential buildings, including offices, schools, medical facilities, private homes, and stores.

## ENVIRONMENTAL BENEFITS

The key areas measured in the LEED certification process reflect the environmental benefits of green building. Sustainable site development involves preserving natural resources for future generations and can include reusing existing buildings, planting around buildings, roof gardens, and underground or earth shelters. Building for water savings and efficiency involves monitoring water supplies and usage, recycling gray or previously used water, and constructing rainwater catchment systems. To improve energy and atmosphere efficiency, buildings can use geographically and climatically appropriate energy resources, including renewable energy. Eliminating chlorofluorocarbons in heating, ventilation, air-conditioning, and refrigeration systems helps reduce ozone depletion. Efforts to conserve materials and resources include using renewable, recycled, local, chemical-free, nonpolluting, and durable materials. Indoor environmental quality can be improved through the use of nontoxic materials, adequate ventilation, temperature controls, and materials that emit few or no volatile organic compounds.

In the twenty-first century, as environmental knowledge and building technologies continue to improve, the green building movement is gaining worldwide momentum. Given that buildings consume more than one-half of the world's resources and generate more than 30 percent of greenhouse gas emissions, the benefits of green building have become increasingly obvious. By 2010 forty-one countries had developed their own LEED initiatives, including Australia, Brazil, Canada, France, India, Israel, Mexico, the United Arab Emirates, and the United Kingdom.

*Alice Myers*

FURTHER READING

Fisanick, Christina, ed. *Eco-rchitecture*. Detroit: Greenhaven Press, 2008.

GreenSource. *Emerald Architecture: Case Studies in Green Building*. New York: McGraw-Hill, 2008.

Johnston, David, and Scott Gibson. *Green from the Ground Up: A Builder's Guide—Sustainable, Healthy, and Energy-Efficient Home Construction*. Newtown, Conn.: Taunton Press, 2008.

Yudelson, Jerry. *The Green Building Revolution*. Washington, D.C.: Island Press, 2008.

# Greenhouse effect

CATEGORY: Weather and climate

DEFINITION: Natural process of atmospheric warming in which solar energy absorbed by the earth's surface is reradiated and absorbed by certain atmospheric gases, primarily carbon dioxide and water vapor

SIGNIFICANCE: Without the natural warming process provided by the greenhouse effect, the earth's atmosphere would be too cold for the planet to support life. However, an intensification of the greenhouse effect resulting from human-generated greenhouse gas emissions could significantly alter surface atmospheric temperatures on a global scale, which would have major environmental and societal implications.

Since 1880, at which point historical measurement records become reliable enough and of sufficient spatial distribution to provide an overview of recent global climate trends, the earth's surface atmospheric temperatures have on average become warmer. During this period—and since the mid-eighteenth century, when the Industrial Revolution began—human activity has released increasing quantities of what are known as greenhouse gases into the atmosphere. These gases include naturally occurring substances such as carbon dioxide, methane, nitrous oxide, and ozone, as well as synthetic chemicals such as chlorofluorocarbons (CFCs), hydrofluorocarbons (HFCs), perfluorocarbons (PFCs), and sulfur hexafluoride. (Water vapor is the most abundant of the naturally occurring greenhouse gases, but human activity has an insignificant influence on its atmospheric concentrations.) Anthropogenic (human-caused) greenhouse gases have been identified as likely contributors to the rise in global surface temperature.

The temperature increase may lead to drastic changes in climate and food production, as well as widespread coastal flooding. As a result, many scientists, organizations, and governments have called for curbs on greenhouse gas emissions. Because their predictions are not definite, however, debate continues about the financial and societal costs of reducing the production of these gases given the lack of certainty regarding the benefits.

GLOBAL WARMING AND HUMAN INTERFERENCE

The naturally occurring greenhouse effect takes place because the gases that make up the atmosphere are able to absorb only particular wavelengths of energy. The atmosphere is largely transparent to shortwave solar radiation, so sunlight basically passes through the atmosphere to the earth's surface. Some is reflected or absorbed by clouds, some is reflected from the earth's surface, and some is absorbed by dust or the earth's surface. Only small amounts are actually absorbed by the atmosphere. Sunlight therefore contributes very little to the direct heating of the atmosphere. On the other hand, the greenhouse gases are able to absorb longwave, or infrared, radiation from the earth, thereby heating the earth's atmosphere.

Discussion of the greenhouse effect has been confused by terms that are imprecise and even inaccurate. For example, when the term "greenhouse effect" was coined during the early nineteenth century, the atmosphere was believed to operate in a manner simi-

## The Greenhouse Effect

Clouds and atmospheric gases such as water vapor, carbon dioxide, methane, and nitrous oxide absorb part of the infrared radiation emitted by the earth's surface and reradiate part of it back to the earth. This process effectively reduces the amount of energy escaping to space and is popularly called the "greenhouse effect" because of its role in warming the lower atmosphere. The greenhouse effect has drawn worldwide attention because increasing concentrations of carbon dioxide from the burning of fossil fuels may result in a global warming of the atmosphere.

Scientists know that the greenhouse analogy is incorrect. A greenhouse traps warm air within a glass building where it cannot mix with cooler air outside. In a real greenhouse, the trapping of air is more important in maintaining the temperature than is the trapping of infrared energy. In the atmosphere, air is free to mix and move about.

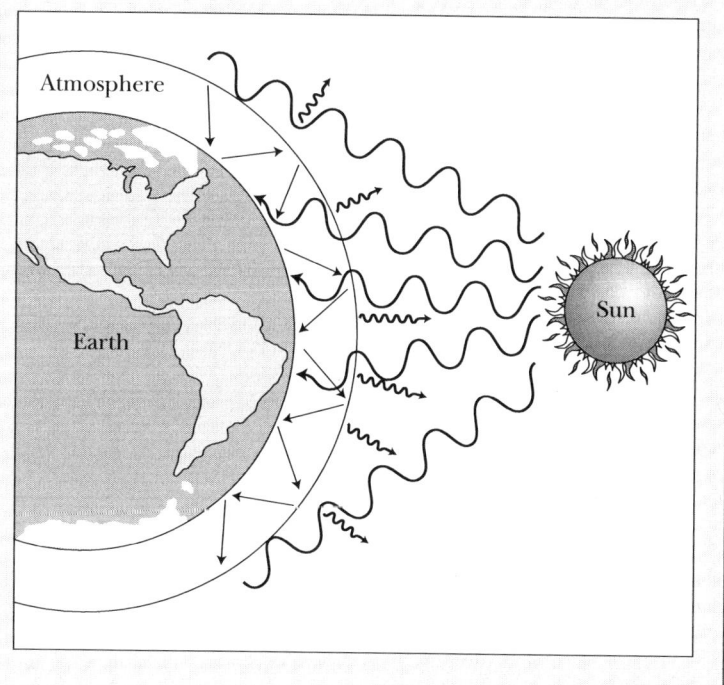

lar to that of a greenhouse, in which glass lets visible solar energy in but is also a barrier preventing the heat energy from leaving. In actuality, the reason that the air remains warmer inside a greenhouse is probably because the glass prevents the warm air from mixing with the cooler outside air. Therefore the greenhouse effect could more accurately be called the "atmospheric effect," but the term "greenhouse effect" continues to be used.

Even though the greenhouse effect is necessary for life on earth, the term gained harmful connotations during the late twentieth century with the discovery of apparently increasing atmospheric temperatures and growing concentrations of greenhouse gases. The concern, however, is not with the greenhouse effect itself, but rather with the intensification or enhancement of the greenhouse effect, presumably caused by increases in the level of gases in the atmosphere resulting from human activity, especially industrialization. Thus the term "global warming" or "global climate change" is a more precise description of this presumed phenomenon.

A variety of human activities have contributed to greenhouse gas emissions. Industry emits a host of natural and synthetic greenhouse gases. Fossil-fuel combustion releases massive quantities of carbon dioxide. Agriculture also affects atmospheric carbon dioxide concentrations, because once natural vegetation and forests have been cleared for agriculture, the crops that replace them may not be as efficient in absorbing carbon dioxide. Increased numbers of ruminant livestock such as cattle and sheep have led to growing levels of methane, which is produced in the animals' digestive systems.

Several greenhouse gases, including CFCs and nitrous oxides, have also been implicated in ozone depletion. Stratospheric ozone shields the earth from solar ultraviolet (shortwave) radiation; therefore, if the concentration of these ozone-depleting gases continues to increase and the ozone shield is depleted, the amount of solar radiation reaching the earth's surface should increase. More solar energy would thus be intercepted by the earth's surface to be reradiated as longwave radiation, which would presumably increase the temperature of the atmosphere.

However, whether there is a direct cause-and-effect relationship between increases in greenhouse gases and surface temperature may be impossible to determine because the atmosphere's temperature has fluctuated widely over millions of years. Over the past 800,000 years, the earth has had several long periods of cold temperatures—during which thick ice sheets covered large portions of the planet's surface—interspersed with shorter warm periods. Since the most recent retreat of the glaciers around 10,000 years ago, the earth has been relatively warm.

## PROBLEMS OF PREDICTION

How much the temperature of the earth might rise as a result of an intensified greenhouse effect is not clear. So far, the temperature increase of around half a degree Celsius (a single degree Fahrenheit) since the late nineteenth century is within the range of normal (historical) trends. The possibility of global warming became a serious concern during the late twentieth century because the decades of the 1980's and the 1990's included some of the hottest years recorded for more than one hundred years. This trend continued into the twenty-first century, with the decade from 2000 to 2009 establishing itself as the warmest on record. On the other hand, warming has not been consistent since 1880. For example, there was a slight global cooling from the 1940's to the 1970's, possibly a result of the increase of another product of fossil-fuel combustion, sulfur dioxide aerosols, which reflect sunlight and thus lessen the amount of solar energy entering the atmosphere. The reduction in sulfur dioxide emissions following the implementation of pollution controls after 1970 may account for the subsequent observed rise in temperature. Similarly, during the early 1990's temperatures declined, most likely because of the ash and sulfur dioxide ejected into the atmosphere during the 1991 eruption of Mount Pinatubo (the second-largest terrestrial volcanic eruption of the twentieth century). By the late 1990's temperatures were on the rise again, suggesting that products of volcanic explosions may have masked the process of global warming.

Proper analysis of global warming is dependent on the collection of accurate temperature records from many locations around the world and over many years. Because human error is always possible, "official" temperature data may not be accurate. This possibility of inaccuracy compromises examination of past trends and predictions for the future. However, the use of satellites to monitor temperatures has served to increase the reliability, comparability, and spatial distribution of the data.

Predictions for the future are hampered in various ways, including lack of knowledge about all the factors and complex interactions affecting atmospheric temperature. Therefore, computer programs cannot be made sufficiently precise to yield accurate predictions. A prime example is the relationship between ocean temperature and the atmosphere. As the temperature of the atmosphere increases, the oceans should absorb much of that heat. Therefore, the atmosphere might not warm as quickly as predicted. However, the carbon dioxide absorption capacity of oceans declines with temperature. Therefore, the oceans would be unable to absorb as much carbon dioxide as before, but exactly how much is unknown. Increased ocean temperatures might also lead to more plant growth, including phytoplankton. These plants would be expected to absorb carbon dioxide through photosynthesis. A warmer atmosphere could hold more water vapor, resulting in the potential for more clouds and more precipitation. Whether that precipitation would fall as snow or rain and where it would fall could also affect air temperatures. Air temperature could drop, as more clouds might reflect more sunlight; on the other hand, more clouds might absorb more infrared radiation.

To complicate matters, any change in temperature would probably not be uniform over the globe. Because land heats up more quickly than water, the Northern Hemisphere, with its much larger landmasses, would be likely to experience greater temperature increases than the Southern Hemisphere. Similarly, ocean currents might change in both direction and temperature. These changes would affect air temperatures as well. In reflection of these complications, computer models of temperature change range widely in their estimates. The Intergovernmental Panel on Climate Change (IPCC) has projected a global temperature increase by the year 2100 of 2 to 3 degrees Celsius (3.6 to 5.4 degrees Fahrenheit).

## MITIGATION ATTEMPTS

International conferences have been held and international organizations have been established to research and minimize the potential detriments of global warming. In 1988 the United Nations Environment Programme and the World Meteorological Organization established the IPCC to assess and compile climate change information for use by policy makers. The IPCC issued assessment reports in 1990, 1995, 2001, and 2007; its fifth assessment report is due in

2013 or 2014. The first of the IPCC's assessment reports concluded that global warming was sure to result if human emissions of greenhouse gases were not brought under control.

In June, 1992, the United Nations Conference on Environment and Development, also known as the Earth Summit, was held in Brazil. Participants devised the United Nations Framework Convention on Climate Change, considered a landmark international treaty. It required signatories to reduce and monitor their greenhouse gas emissions. Developed nations agreed on a voluntary year 2000 target of stabilizing their emissions at 1990 levels, a goal that many ratifying governments failed to meet.

A binding agreement, the Kyoto Protocol, was adopted in December, 1997, and entered into force in February, 2005. It set compulsory targets for reducing emissions of carbon dioxide, methane, nitrous oxide, sulfur hexafluoride, HFCs, and PFCs over the five-year period of 2008 through 2012 for thirty-seven industrialized countries and the European Community; no emissions targets were established for developing countries. The treaty allows signatory nations to use afforestation (the creation of forests where none existed before) to offset emissions and includes the economic incentive of allowing the trading of emissions permits—that is, countries that meet their targets can sell their excess permits to other nations. The United States, which produces roughly one-fourth of the world's greenhouse gas emissions, has not ratified the treaty. President George W. Bush formally withdrew the nation from the protocol in 2001, citing the damage it would do to the American economy and stating his objection to the exemption of developing nations such as India and China—both fast-growing polluters—from obligations to make emissions cuts before 2012.

The first commitment period of the Kyoto Protocol ends in 2012, by which time a follow-up international framework must be in place. Conferences on post-Kyoto climate policy have proved contentious. Attempts to negotiate a successor to the protocol, notably at the United Nations Climate Change Conference held in Copenhagen, Denmark, in December, 2009, generated more than one hundred domestic policies and government initiatives by participating nations to reduce greenhouse gas emissions but failed to produce a new international treaty.

*Margaret F. Boorstein*
*Updated by Karen N. Kähler*

FURTHER READING

Gore, Al. *An Inconvenient Truth: The Planetary Emergency of Global Warming and What We Can Do About It.* New York: Rodale Books, 2006.

Houghton, John Theodore. *Global Warming: The Complete Briefing.* 4th ed. New York: Cambridge University Press, 2010.

Lankford, Ronald D., ed. *Greenhouse Gases.* Detroit: Greenhaven Press, 2009.

Schneider, Stephen Henry, et al., eds. *Climate Change Science and Policy.* Washington, D.C.: Island Press, 2010.

Shulk, Bernard F., ed. *Greenhouse Gases and Their Impact.* New York: Nova Science, 2007.

Weart, Spencer W. *The Discovery of Global Warming.* Rev. ed. Cambridge, Mass.: Harvard University Press, 2008.

# Greenhouse gases

CATEGORIES: Atmosphere and air pollution; pollutants and toxins

DEFINITION: Atmospheric gases that allow sunlight to reach the earth's surface but at least partially block infrared from radiating back into space

SIGNIFICANCE: Greenhouse gases raise the earth's average temperature 33 degrees Celsius (59 degrees Fahrenheit) above what it would be if these gases were not present in the atmosphere. Most scientists agree that human activities are contributing to increased concentrations of greenhouse gases and thus to increases in the average surface temperature of the earth.

The majority of scientists accept that a small global warming has taken place—that is, the earth's surface temperature has warmed about 0.7 degree Celsius (1.3 degrees Fahrenheit)—since the end of the nineteenth century. At least half of this temperature rise is attributed to the release of greenhouse gases into the atmosphere by human beings. It is thought that if greenhouse gases were returned to their 1990 levels, the temperature would still rise another 0.5 degree Celsius (0.9 degree Fahrenheit). Since greenhouse gases are transparent to visible light, sunlight passes through the atmosphere and warms the earth's surface. The warmed surface radiates infrared into the sky, where greenhouse gases absorb infrared and then

## U.S. Greenhouse Gas Emissions
### (millions of metric tons)

|                | 1990    | 2000    | 2002    | 2003    | 2004    | 2005    | 2006    |
|----------------|---------|---------|---------|---------|---------|---------|---------|
| Carbon dioxide | 5,017.5 | 5,890.5 | 5,875.9 | 5,940.4 | 6,019.9 | 6,045.0 | 5,934.4 |
| Methane gas    | 708.4   | 608.0   | 598.6   | 603.7   | 605.9   | 607.3   | 605.1   |
| Nitrous oxide  | 333.7   | 341.9   | 332.5   | 331.7   | 358.3   | 368.0   | 378.9   |
| High GWP gases | 87.1    | 138.0   | 137.8   | 136.6   | 149.4   | 161.2   | 157.6   |

*Source:* U.S. Energy Information Administration, *Emissions of Greenhouse Gases in the United States, 2006,* 2006.
*Note:* High GWP (global warming potential) gases are hydrofluorocarbons, perfluorcarbons, and sulfur hexafluoride.

reradiate it. They radiate about half of the infrared upward into space and half of it back down to the earth's surface. Since the earth absorbs more energy than it radiates back into space, it heats up until the energies entering and leaving are equal. Balance is possible because a hotter earth radiates with greater intensity and at shorter wavelengths where greenhouse gases allow more infrared to escape into space.

Water vapor strongly absorbs infrared of about 3 microns wavelength (3,000 nanometers), and so does carbon dioxide. Adding more carbon dioxide will not change the amount of energy passing out into space, since water vapor absorbs all of the energy near that wavelength. However, more carbon dioxide would make a difference at 4.5 microns wavelength (4,500 nanometers) because water vapor does not absorb at that wavelength. This means that doubling the amount of carbon dioxide in the atmosphere does not necessarily double the effect of carbon dioxide. In fact, most climate scientists believe that increasing carbon dioxide will increase the surface temperature, which will cause more water to evaporate. More water vapor in the atmosphere should further warm the earth's surface, but it will also cause more clouds that reflect sunlight back into space. More clouds will cool the earth. Under these competing processes it is believed that temperature will increase until a new equilibrium is reached.

### EARTH'S MAJOR GREENHOUSE GASES

Water vapor, carbon dioxide, methane, nitrous oxide, fluorocarbons, and ozone are the major greenhouse gases. The effect of a particular gas depends on which other gases are present, how much of the gas is present, and how likely a gas molecule is to absorb infrared radiation.

Water vapor is the most important greenhouse gas, but human activities have no direct effect on the global average amount of water vapor, which is fixed mainly by evaporation from the earth's oceans. Other greenhouse gases are significantly affected by human activities. Burning fossil fuels and deforestation in the Tropics increase the amount of carbon dioxide in the atmosphere. Rice paddy farming and the digestive processes of livestock produce large amounts of methane. Fertilizers used in farming produce nitrous oxide, and refrigeration systems and some manufacturing processes release chlorofluorocarbons, other perfluorocarbons, and sulfur hexafluoride.

Since the mid-nineteenth century, carbon dioxide concentration in the air has increased from 280 parts per million to almost 380 parts per million. Normal carbon is carbon 12 (6 protons and 6 neutrons in the nucleus), but about 1 percent of carbon is carbon 13 (6 protons and 7 neutrons). Plants prefer carbon 12, so plant carbon (and fossil fuels from plants) has a smaller ratio of carbon 13 to carbon 12 than the atmosphere does. Analysis of air bubbles in ice cores shows that the ratio of carbon 13 to carbon 12 in the atmosphere has been decreasing since the mid-nineteenth century, presumably because of increased burning of fossil fuels and the practice of setting fire to forests to clear land.

### POLITICS

Under the sponsorship of the United Nations, the Kyoto Protocol was adopted in December, 1997, and went into force in February, 2005. Seeing the protocol as flawed because it places no limits on China (the world's largest polluter) or India, the United States opted out of the protocol. Under the protocol, industrialized nations set goals for greenhouse gas reductions, and developing nations negotiated for money from those nations to help them industrialize with fewer greenhouse emissions. By 2007 only Germany, Norway, France, and the United Kingdom were meeting their goals. Several nations that had been part of the former Soviet Union reduced emissions, in large part because their economies floundered.

In December, 2009, the U.S. Environmental Protection Agency (EPA) released a finding that greenhouse gases threaten the public health and welfare of the American people. The finding will allow the EPA to regulate greenhouse gas emissions in the United States if the Congress fails to pass legislation to do so.

In November, 2009, a large number of e-mails were stolen from the Climatic Research Unit at the University of East Anglia, United Kingdom. Some climate change skeptics alleged that the e-mails provided evidence that scientists had doctored data on global warming. Subsequent investigations found evidence of frustration on the part of the climate researchers and a lack of willingness among them to share raw data, but no evidence of wrongdoing. This incident nevertheless cast some doubt on the East Anglia data on global warming, despite the fact that the findings of the Climatic Research Unit are supported by a great deal of data from other sources. This doubt probably contributed to the weakness of the agreements reached in regard to greenhouse gas emissions at the United Nations Climate Change Conference held in Copenhagen, Denmark, in December, 2009.

*Charles W. Rogers*

## FURTHER READING

Houghton, John. *Global Warming: The Complete Briefing.* New York: Cambridge University Press, 2009.

Singer, S. Fred, and Dennis T. Avery. *Unstoppable Global Warming: Every 1,500 Years.* Updated ed. Lanham, Md.: Rowman & Littlefield, 2008.

Weart, Spencer W. *The Discovery of Global Warming.* Rev. ed. Cambridge, Mass.: Harvard University Press, 2008.

# Hazardous and toxic substance regulation

CATEGORIES: Human health and the environment; pollutants and toxins

DEFINITION: Legally enforceable rules pertaining to substances that pose a threat to people, animals, or the environment because of their chemical, physical, or biological properties

SIGNIFICANCE: The regulation of hazardous and toxic substances is necessary in order to protect people and the environment from the by-products of industrialization. Policy makers must find a balance between safeguarding against undesirable health or environmental effects and reaping the benefits of the activities that produce such hazards.

The survival of the human species is inseparable from the preservation of the environment. This concern may be divided into three categories: the health of the general population, occupational hazards, and the availability of food and water supplies that are pure and free from pollutants. Within these categories, industrial development has created new problems and environmental pressures, even as technological advances improve the ability to detect, study, and analyze environmental changes.

Since ancient times humans have been concerned with issues surrounding public health. The most common concerns before the Industrial Revolution of the eighteenth and nineteenth centuries were related to food and water supplies. Tasks involving occupational hazards were left for slaves and the lower classes and did not, therefore, receive much recognition. After the Industrial Revolution, contact with occupational hazards and toxic substances became much more commonplace.

By the early twentieth century, conditions had worsened for the average working person, and those not faced with occupational hazards were beginning to feel the effects of an industrialized society that gave rise to hazardous by-products and toxic waste. Consciousness of environmental problems began to increase, but the rate of industrial development outstripped societal awareness of the hazards such development was creating.

Although Congress made attempts to regulate the hazardous and toxic wastes being generated from mining, agriculture, and industry, progress was slow because there was little public interest in the situation. Awareness of the negative impact of industrialization substantially increased during the 1960's as environmental activists such as Barry Commoner and Rachel Carson succeeded in publicizing issues relating to the deterioration of the environment. Public interest reached a peak with the celebration of the first Earth Day in 1970, and Congress began to respond to public pressure by strengthening the laws that had previously been passed to regulate air and water pollution.

Efforts to decrease hazardous and toxic waste levels in the environment were relatively ineffective until August 2, 1978, when New York State officials ordered the emergency evacuation of 239 families living within two blocks of the Love Canal chemical waste dump in Niagara Falls. Headlines throughout the nation declared Love Canal the largest human-made environmental disaster in decades, and Americans began paying attention to the issue of hazardous and toxic wastes. The general public held policy makers accountable for the disaster and the prevention of similar occurrences in the future.

### DEVELOPMENT OF U.S. LEGISLATION

The U.S. regulatory style is considered to be one of open conflict, with interest groups, various media, legislators, and the courts all playing important roles in the development of laws. Policy makers must also rely on the consensus of scientists in understanding the risks and magnitude of problems involving hazardous and toxic materials. This reliance complicates and considerably slows the regulation process. Policy makers must also make tough decisions on the health benefits versus economic costs of controlling hazardous and toxic wastes.

This difficult task of creating legislation that works to protect the environment and the health of citizens in a hazardous society can be broken down into a four-step process: creating a law, putting the law to work, creating a regulation, and enforcing the law. The collaboration of Congress, government agencies, and societal awareness and responsibility is needed to create effective legislation.

To create new legislation, a member of Congress must propose a bill, which, if approved by both the House of Representatives and the Senate and signed by the president, becomes a new law. The act is codified by the House of Representatives and published in the United States Code. To put the law to work, regulations for the law must be created by government agencies authorized by Congress. Regulations are rules about the law, specifying what is legal and what is not. To create a regulation, the authorized government agency, usually the U.S. Environmental Protection Agency (EPA) in the case of hazardous and toxic substances, determines the need to form a regulation. The regulation is proposed on the Federal Register, and members of the public are allowed to provide input in the form of comments and suggested modifications. Revisions to the regulation may be made ac-

cordingly. Once a completed regulation is finished, it is published in the Code of Federal Regulations (CFR). Twice a year each agency publishes a comprehensive report that describes all the regulations it is working on or has recently finished. Laws and regulations are enforced by the government agency that put them into effect.

### IMPORTANT U.S. LEGISLATION

The responsibility for regulating and enforcing hazardous materials laws in the United States is primarily split among a few federal regulatory agencies. There is some overlap between the fields that regulate hazardous and toxic materials. Federal governmental agencies that are involved in regulating hazardous and toxic substances are the EPA, the Occupational Safety and Health Administration (OSHA), and the Department of Transportation (DOT).

Several environmental laws affect hazardous and toxic materials policy in the United States. The Occupational Safety and Health Act, passed in 1970, provides standards of allowable exposure to toxic chemicals in the workplace. This law establishes occupational exposure limits for hundreds of toxic and hazardous substances. The act also establishes labeling standards for equipment, standards for personal protection, and monitoring requirements for the health of workers. With passage of the act, OSHA and the National Institute of Occupational Safety and Health (NIOSH) were created. The 1972 amendments to the Federal Insecticide, Fungicide, and Rodenticide Act (FIFRA) established a regulatory program for the EPA to control the manufacture of potentially harmful pesticides. This legislation was created to prevent the adverse environmental effects posed by new pesticides and to ensure safety standards for people using pesticides.

The Safe Drinking Water Act (SDWA) of 1974 was established to protect groundwater and drinking-water sources from contamination by hazardous chemicals. The act sets two levels of standards to limit the amount of contamination that might be found in drinking water: primary standards with a maximum contaminant level (MCL) to protect human health and secondary standards that relate to color, taste, smell, and other physical characteristics. The Hazardous Materials Transportation Act (HMTA) of 1975 provides a high level of environmental protection during hazardous waste transportation managed by the DOT. By requiring special packing and routing, the act ensures the careful shipment of hazardous substances.

The 1976 Resource Conservation and Recovery Act (RCRA) and its amendments deal with the ongoing management of solid wastes throughout the United States. With RCRA, a "cradle-to-grave" approach to hazardous waste management was introduced that was designed to protect groundwater supplies by focusing on the treatment, storage, and disposal of such wastes. RCRA focuses on five main areas for hazardous waste management: identification and classification of hazardous waste; requirements for generators of hazardous waste to identify themselves so that hazardous waste activities can be tracked and standards of operation for generators can be established; adoption of standards for the transportation of hazardous wastes; standardization of treatment, storage, and disposal facilities; and provisions for enforcement of the standards with a program of legal penalties for noncompliance. RCRA classifies waste materials based on four characteristic properties: ignitability, corrosivity, reactivity, and toxicity. The need to clean up abandoned toxic waste sites such as Love Canal, which RCRA did not address, gave rise to the Comprehensive Environmental Response, Compensation, and Liability Act (CERCLA) of 1980, widely known as Superfund, and its amendments, including the Superfund Amendments and Reauthorization Act (SARA) of 1986.

The Toxic Substances Control Act (TSCA) of 1976 requires that all chemicals produced in or imported into the United States be tested, regulated, and screened for toxic effects prior to commercial manufacture. This law bans the manufacture of polychlorinated biphenyls (PCBs) and regulates asbestos. The EPA works with other federal agencies under this law to fill in the gaps of the other acts that attempt to manage hazardous materials. Additional laws under which

## Ford Signs the Toxic Substances Control Act

*On October 12, 1976, President Gerald R. Ford made this statement before signing the Toxic Substances Control Act into law:*

I believe this legislation may be one of the most important pieces of environmental legislation that has been enacted by the Congress.

This toxic substances control legislation provides broad authority to regulate any of the tens of thousands of chemicals in commerce. Only a few of these chemicals have been tested for their long-term effects on human health or the environment. Through the testing and reporting requirements of the law, our understanding of these chemicals should be greatly enhanced. If a chemical is found to present a danger to health or the environment, appropriate regulatory action can be taken before it is too late to undo the damage.

The legislation provides that the Federal Government, through the Environmental Protection Agency, may require the testing of selected new chemicals prior to their production to determine if they will pose a risk to health or the environment. Manufacturers of all selected new chemicals will be required to notify the Agency at least 90 days before commencing commercial production. The Agency may promulgate regulations or go into court to restrict the production or use of a chemical or to even ban it if such drastic action is necessary.

The bill closes a gap in our current array of laws to protect the health of our people and the environment. The Clean Air Act and the Water Pollution Control Act protect the air and water from toxic contaminants. The Food and Drug Act and the Safe Drinking Water Act are used to protect the food we eat and the water we drink against hazardous contaminants. Other provisions of existing laws protect the health and the environment against other polluting contaminants such as pesticides and radiation. However, none of the existing statutes provide comprehensive protection. . . .

The administration, the majority and minority members of the Congress, the chemical industry, labor, consumer, environmental, and other groups all have contributed to the bill as it has finally been enacted. It is a strong bill and will be administered in a way which focuses on the most critical environmental problems not covered by existing legislation while not overburdening either the regulatory agency, the regulated industry, or the American people.

the EPA acts include the Clean Air Act (CAA) and amendments and the Clean Water Act (CWA) and amendments.

### INTERNATIONAL REGULATORY EFFORTS

A host of international agreements and guidelines shape national and regional hazardous and toxic substances regulations around the world. The Basel Convention on the Control of Transboundary Movements of Hazardous Wastes and Their Disposal, for example, was adopted in 1989 to keep developed countries from dumping their hazardous wastes in developing countries. It addresses the transport of hazardous wastes among countries and illegalizes such transport without prior informed consent. Similarly, the Waigani Convention to Ban the Importation into Forum Countries of Hazardous and Radioactive Wastes and to Control the Transboundary Movements and Management of Hazardous Wastes Within the South Pacific Region was adopted in 1995 to halt the practice of waste traders of using the South Pacific region as a dump for hazardous and nuclear wastes. The convention also provides for the environmentally responsible management and disposal of existing wastes in this region.

The Rotterdam Convention on the Prior Informed Consent Procedure for Certain Hazardous Chemicals and Pesticides in International Trade, adopted in 1998, facilitates the environmentally sound management of severely restricted pesticide formulations and other extremely hazardous chemicals by requiring countries to share information regarding these substances. Exporters must supply information to importing countries and must abide by the importers' wishes. The Stockholm Convention on Persistent Organic Pollutants, adopted in 2001, restricts and ultimately prohibits the production, use, import, and export of bioaccumulative and environmentally persistent chemicals such as PCBs, furans, dioxins, and dichloro-diphenyl-trichloroethane (DDT). The Strategic Approach to International Chemicals Management, a policy framework designed to foster safe practices in chemical production and use worldwide, was adopted in 2006.

To minimize the likelihood of hazardous and toxic materials transport accidents that could harm people or property or damage the environment, the United Nations Economic and Social Council developed the U.N. Recommendations on the Transport of Dangerous Goods. The council administers regional agreements regarding hazardous and toxic materials transport and works to keep various nations' regulatory systems from impeding the flow of trade. Following the 1992 Earth Summit, at which the harmonization of chemical classification and labeling was identified as an international priority, the Globally Harmonized System of Classification and Labeling of Chemicals (GHS) was created to facilitate international trade, transport, and management of hazardous and toxic substances. Various modes of transport have their own regulatory schemes, notably the International Air Transport Association Dangerous Goods Regulations, the International Maritime Dangerous Goods Code, and the Regulations Concerning the International Carriage of Dangerous Goods by Rail.

*Marcie L. Wingfield and Massimo D. Bezoari*
*Updated by Karen N. Kähler*

FURTHER READING

Bergeson, Lynn L. *TSCA: The Toxic Substances Control Act.* Chicago: American Bar Association, 2000.

Cranor, Carl F. *Toxic Torts: Science, Law, and the Possibility of Justice.* New York: Cambridge University Press, 2006.

Griffin, Roger D. *Principles of Hazardous Materials Management.* 2d ed. Boca Raton, Fla.: CRC Press, 2009.

Hill, Marquita K. "Hazardous Waste." In *Understanding Environmental Pollution.* 3d ed. New York: Cambridge University Press, 2010.

LaGrega, Michael D., Philip L. Buckingham, and Jeffrey C. Evans. *Hazardous Waste Management.* 2d ed. New York: McGraw-Hill, 2001.

Sprankling, John G., and Gregory S. Weber. *The Law of Hazardous Wastes and Toxic Substances in a Nutshell.* 2d ed. St. Paul, Minn.: Thomson/West, 2007.

Switzer, Carole Stern, and Peter L. Gray. *CERCLA: Comprehensive Environmental Response, Compensation, and Liability Act (Superfund).* 2d ed. Chicago: American Bar Association, 2008.

# Hiroshima and Nagasaki bombings

CATEGORIES: Disasters; nuclear power and radiation

THE EVENTS: The dropping of two atomic bombs on cities in Japan during World War II—a uranium bomb on Hiroshima and a plutonium bomb on Nagasaki

DATES: August 6 and 9, 1945

SIGNIFICANCE: The use of atomic weapons by the United States is credited with ending the war with Japan, but debates continue regarding the necessity and moral justification of the bombings. The bomb blasts and the radiation emitted immediately on detonation were responsible for widespread devastation and the deaths of hundreds of thousands of people, but the effects of radiation on the surrounding areas were less long-lasting than had been expected.

By the summer of 1945, Japan had lost the war against the United States and the other Allied Powers, but the ruling Japanese war cabinet was deadlocked—three for and three against—on a decision on surrender. With the tie vote, the previous policy of war prevailed. The Japanese devised a strategy to end the war that would force the Allies to invade the Japanese homeland, where so many would be killed that Japan would receive more favorable surrender terms to end the war. The Allies did not invade Japan, however; on August 6 and 9, 1945, the United States dropped atomic bombs on the cities of Hiroshima and Nagasaki. Even after the devastation caused by the bombings, the war cabinet remained deadlocked, but Emperor Hirohito survived an attempted coup and gained enough support that he was able to lead Japan's surrender.

## THE BOMBS

The bomb code-named Little Boy exploded 580 meters (1,900 feet) over Hiroshima. It weighed 4,400 kilograms (9,700 pounds). Its great power came from 64 kilograms (141 pounds) of uranium 235. The uranium was assembled into critical mass (the amount needed to sustain a chain reaction) in 1 millisecond by the firing of a 25-kilogram (55-pound) uranium bullet from a short cannon into 39 kilograms (86 pounds) of uranium shaped into three target rings. Although the bomb was only 1.3 percent efficient, it exploded with the power of 13 kilotons of trinitrotoluene (TNT).

A three-second thermal pulse from the blast burned the exposed skin of people as far away as 3.5 kilometers (2.2 miles) from ground zero (the point directly below the burst) and started fires throughout the city. The blast crushed buildings around ground zero, leaving only skeletons of reinforced concrete or steel. Numerous fires spread, joined, and formed a violent firestorm that consumed everything combustible within 13 square kilometers (5 square miles). It is estimated that some 70,000 people were killed, and another 100,000 died over the next five years from bomb-related injuries. An estimated 60 percent of the deaths were from burns, 30 percent from the blast, and 10 percent from radiation.

Because the Japanese did not surrender immediately, and to demonstrate that the Hiroshima bomb was not a fluke, the United States detonated a plutonium bomb, code-named Fat Man, 595 meters (1,952 feet) over Nagasaki three days later. (The bombs'

code names, Little Boy and Fat Man, referred to U.S. president Harry Truman and British prime minister Winston Churchill, respectively.) Fat Man weighed 4,900 kilograms (10,800 pounds), half of which was high explosives used to compress the softball-sized plutonium core into critical mass. The 22-kiloton blast and subsequent fire destroyed 6.7 square kilometers (2.6 square miles). The destruction was not greater because part of the city was protected by a hill. Approximately 70,000 people were killed, and another 70,000 died over the next five years from bomb-related injuries. An estimated 77 percent of these deaths were from burns, 16 percent from the blast, and 7 percent from radiation.

## RADIATION

Prompt radiation consists of the neutrons and gamma rays emitted as a nuclear bomb explodes and during the first minute thereafter. About 85 percent of the people within 1 kilometer (0.6 mile) of ground

*A mushroom cloud forms over the Japanese city of Nagasaki on August 9, 1945, moments after the United States dropped the second atomic bomb ever used in warfare.* (National Archives)

zero of both blasts were killed outright by blast or heat, or they died within the first year of radiation or injuries. People up to 2 kilometers (1.2 miles) away received significant but nonlethal prompt doses of radiation. A study of thirty women who had been pregnant when they were exposed to radiation within 2 kilometers of ground zero and showed signs of radiation damage found that only sixteen of the children they subsequently bore lived more than one year, and four of those children were developmentally disabled. Deaths from prompt radiation began during the first week and peaked three to four weeks later. Between 1950 and 1990 an estimated 850 people who had been in the vicinities of the Hiroshima and Nagasaki bombings died of radiation-caused cancer, and another 400 to 800 died of radiation-induced noncancer injuries.

Radiation after the first minute of a nuclear bomb blast is called delayed radiation. Near ground zero, neutrons make elements such as sodium, aluminum, and manganese radioactive, but these newborn isotopes have short half-lives and disappear within minutes to hours. The other source of delayed radiation is fallout. It consists of radioactive bomb vapor that has condensed into small, solid particles. Had the Hiroshima and Nagasaki bombs been ground bursts, dirt and rocks would have been sucked up into the fireball and made radioactive by neutrons, which would have made the fallout much worse. Since the bombs were air bursts, this did not happen. Within an hour of each burst, soot from the fires helped precipitate black rain. In spite of reports to the contrary, the rain was only mildly radioactive. Direct measurements found the highest levels to be in the Nishiyama district of Nagasaki, but even there radiation levels should not have caused lasting harm. People who came into the city right after the explosion and were exposed to the fallout and residual radiation showed no symptoms attributable to radiation.

Although some scientists expected that Hiroshima and Nagasaki would be uninhabitable for generations because of the radiation caused by the atomic weapons, in reality radiation in the cities had returned to near background level one month after the bombings. Both cities were rebuilt and achieved their prebombing populations within ten years. In the twenty-first century they are modern and prosperous.

*Charles W. Rogers*

FURTHER READING
Pike, Frances. *Empires at War: A Short History of Modern Asia.* London: I. B. Tauris, 2010.
Rotter, Andrew J. *Hiroshima: The World's Bomb.* New York: Oxford University Press, 2008.
Schull, William J. *Effects of Atomic Radiation: A Half-Century of Studies from Hiroshima and Nagasaki.* New York: Wiley-Liss, 1995.

## Incineration of waste products

CATEGORY: Waste and waste management
DEFINITION: Burning of waste materials under controlled conditions
SIGNIFICANCE: Incineration has several benefits as a method of waste disposal: It reduces the volume of waste by about 95 percent while producing useful amounts of heat, and it can be used to sterilize medical waste and to neutralize dangerous chemicals. Waste incineration can also produce environmentally harmful by-products if it is not conducted carefully, and for this reason the practice has not achieved widespread acceptance in the United States.

The incinerators used in burning waste products vary in type depending on the kinds and amounts of wastes to be processed. Solid household waste usually generates a lot of ash; if large amounts of such waste are incinerated, ash must be continuously extracted. This may be done with a moving grate incinerator, in which the grate is a conveyor belt. Waste is dumped onto the grate's front end, and ash and clinkers (unburned solids) are removed at the back end. Air is forced up through the grate to cool the grate and to aid combustion. If necessary, the grate can also be water-cooled. Air is also injected above the grate to ensure complete combustion of the gases. European law concerning waste incineration requires that the gases reach at least 850 degrees Celsius (1,560 degrees Fahrenheit) for at least 2 seconds to guarantee the breakdown of toxic organic material. If the gases are not hot enough, an oil burner is used, so wastes with relatively low fuel value can be treated in a moving grate incinerator.

If heat from the incinerator is to be used, the combustion gases pass through a heat exchanger. The heated working fluid from the exchanger may be used to produce steam that powers a turbine and produces electricity, or the hot fluid may carry energy elsewhere. Among the nations that have found ways to use the heat created by incinerators is Denmark; in 2005 waste incinerators produced 4.8 percent of the electricity and 13.7 percent of the space heating used in that country. A typical moving grate incinerator can handle 38.5 tons of municipal waste per hour, and 1.1 tons of such waste may produce 0.67 megawatt-hours of electricity and 2 megawatt-hours of space heating.

Pollutants not destroyed by heat or burning during incineration may be confined to the ash, which must be disposed of properly. Pollutants in the flue gas must be neutralized. Ash mixed with flue gas, called fly ash, must be captured. The cheapest capture method works best on larger particles: Flue gas is made to swirl in a cyclone chamber so that centrifugal force drives dust particles toward the outer chamber wall, from which they then drop into a hopper. In electrostatic precipitation, dust particles are given a negative charge and then the flue gas is passed between large, highly positively charged plates. The charged dust particles migrate to the plates and stick to them. Periodically shaking the plates allows the dust to fall down into a hopper for later removal. An excellent precipitator can remove 99.9 percent of fly ash.

Another capture method involves forcing flue gas through filter bags in a bag room. A final method involves spraying water droplets into the gas and letting the drops gather the dust as they fall. If there are pollutants in the flue gas, chemicals can be added to the water spray, or they may be blown in as a dry powder. Acids are neutralized with sodium bicarbonate, and bases and chlorides are treated with sodium hydroxide.

## OTHER TYPES OF INCINERATORS

Medical wastes must be incinerated at more than 1,000 degrees Celsius (1,832 degrees Fahrenheit) to ensure the destruction of all pathogens. Such wastes are burned in a rotary kiln incinerator, which is completely lined with firebricks and has no moving parts in the heated region, so it can withstand such temperatures. The incinerator consists of a rotating horizontal cylinder with one end higher than the other. Waste is put in at the high end, and the burning residue automatically migrates to the lower end for ash removal.

When the waste particles and their composition are relatively homogeneous, as they are in liquid municipal sludge waste, a fluidized bed incinerator is used. The bed is a layer of sand about 1 meter (3.3

feet) thick. Air blown from beneath the sand bed lifts the sand particles and keeps them suspended. Fuel is blown in and ignited at start-up, then waste is sprayed onto the fluidized bed. Fluid waste droplets are quickly reduced to particles that are burned. They are given sufficient oxygen and uniformly heated from all sides so that burning is complete. Flue gases pass through a heat exchanger, preheating the air blown from under the sand bed. The gases are treated as necessary before they are vented to the atmosphere.

In a plasma arc incinerator, a plasma arc (which looks like a very big spark) is created by the passing of a high-voltage, high-amperage current between two electrodes. A gas passing between the electrodes is heated to 13,900 degrees Celsius (25,000 degrees Fahrenheit). The gas temperature drops to 2,800 to 4,400 degrees Celsius (5,100 to 8,000 degrees Fahrenheit) as it circulates into the incinerator's waste-containing chamber. The temperature is hot enough to break chemical compounds into their constituent atoms, so with proper flue gas treatment a plasma arc incinerator can handle difficult wastes such as batteries and asbestos.

Incineration provides one of the best methods for destroying dangerous chemicals such as those used in chemical weapons. Since the United Nations Convention on the Prohibition of the Development, Production, Stockpiling, and Use of Chemical Weapons and on Their Destruction went into effect in 1997, the United States has been using incineration to destroy its stockpile of some 33,000 tons of nerve gas and mustard gas. The poison chemicals are broken down by extremely high temperatures; they are heated in an incinerator to 1,500 degrees Celsius (2,700 degrees Fahrenheit), and their former containers are heated to 900 degrees Celsius (1,650 degrees Fahrenheit) for 2.5 hours. By 2010 the United States had destroyed 75 percent of its chemical weapons stockpile.

*Charles W. Rogers*

## FURTHER READING

Pichtel, John. "Incineration of MSW." In *Waste Management Practices: Municipal, Hazardous, and Industrial.* Boca Raton, Fla.: CRC Press, 2005.

Royte, Elizabeth. *Garbage Land: On the Secret Trail of Trash.* New York: Little, Brown, 2005.

Tammemagi, Hans. "Incineration: The Burning Issue." In *The Waste Crisis: Landfills, Incinerators, and the Search for a Sustainable Future.* New York: Oxford University Press, 1999.

# Indoor air pollution

CATEGORIES: Atmosphere and air pollution; human health and the environment

DEFINITION: Contamination of the air contained within buildings

SIGNIFICANCE: Pollutants in indoor air are believed to cause thousands of deaths each year—mainly from lung cancer caused by radon and carbon monoxide poisoning—as well as a considerable amount of illness and discomfort.

Because most people in developed nations spend more time indoors than outdoors, the air quality in homes, offices, stores, and other buildings can have a greater effect on human health than the quality of outdoor air. The term "sick building syndrome" is used when a majority of a building's occupants experience health and comfort problems caused by a variety of indoor pollutants that are difficult to identify.

Outdoor air is one source of indoor air pollution, because the ventilation systems bring in air from the outside. Fortunately, some pollutants are trapped as they enter buildings; particulate matter, for example, may stick to walls and pass no further than entryways.

## CARBON MONOXIDE AND CARBON DIOXIDE

Far more important to indoor air are activities that occur inside buildings. One significant contributor to indoor air pollution is combustion; such pollution may come from appliances or from tobacco-smoking building occupants. An unvented or improperly vented furnace or water heater, or a cracked heat exchanger in a furnace, may allow combustion products such as carbon monoxide to enter the indoor space. Carbon monoxide, which can also enter buildings from garages in which motor vehicle engines are running, causes hundreds of accidental deaths annually in the United States and produces a large amount of often-unrecognized illness. Tobacco smoke is known to be harmful not only to smokers but also to nonsmokers exposed to a significant amount of secondhand smoke; tobacco smoke is also a major source of annoyance and discomfort to most people.

Carbon dioxide, normally regarded as nontoxic, causes nausea and headaches at elevated levels and should not exceed 5,000 parts per million (ppm). Even at 1,000 ppm, however, a buildup of carbon dioxide will make a room seem stuffy. Unvented gas or kerosene space heaters should never be used indoors because they necessarily lead to high carbon dioxide levels and often produce high levels of carbon monoxide, nitrogen oxide, and sulfur oxide (the last of these is a particularly severe problem with kerosene heaters, because kerosene contains sulfur).

## BUILDING MATERIALS AND VOLATILE COMPOUNDS

Building materials can also contribute to indoor air pollution. Asbestos, a known cause of lung cancer, may be present in indoor air if insulation or other materials containing asbestos have broken down; these can be especially dangerous when the materials are being removed or otherwise disturbed, such as during building renovation. Formaldehyde, a major component of urea-formaldehyde foam insulation, particle board, and some packaging materials, can produce acute eye, nose, and throat irritation at levels below 1 ppm. Formaldehyde is mainly a concern during the first few years after building construction or renovation, after which it eventually disappears.

A variety of other volatile organic compounds may contribute to poor indoor air quality. These include acetone and other ketones, alcohols, aromatic hydrocarbons (such as benzene and toluene), and halogenated hydrocarbons (such as methylene chloride) found in adhesives, household cleaners, enamels, glues, paints, solvents, and varnishes. Indoor hobbies or renovations involving such compounds should be undertaken only in areas that are well ventilated.

Microorganisms such as bacteria, fungi, molds, and viruses can be dangerous in buildings that are not kept clean. They appear to be most troublesome in buildings with low relative humidity, but some thrive in moist areas. The organism responsible for Legionnaires' disease has been found capable of growing in poorly maintained cooling and ventilation systems.

## VENTILATION

The ventilation rate of a building plays an important role in its indoor air quality. The concentration of air pollutants can rise at a rapid rate in a poorly ventilated building because the pollutants generated inside the building are not being removed quickly enough. This problem has been aggravated since the energy crisis of the 1970's by the practice of making buildings airtight in order to reduce energy costs associated with the heating or cooling of outdoor air that has entered a building. It is possible for a building to be ventilated well and yet have low energy costs if an air-to-air heat exchanger is used; this device allows the incoming and outgoing airstreams to pass near each other across a thin conducting barrier, so that a large fraction of the heat from the outgoing air in winter is transferred to the incoming air. On the other hand, a building in which there is little indoor generation of air pollutants can be airtight and still have superior air quality; increasing the ventilation rate in such a building may actually decrease indoor air quality by bringing in outdoor pollution.

Radon levels are particularly and subtly dependent on the way a building is ventilated. Radon is normally present in underground air in concentrations sufficient to cause concern if even a small percentage of the air in the building comes from underground; the ventilation of the lower level of a building may actually increase this percentage.

*Laurent Hodges*

## FURTHER READING

Burroughs, H. E., and Shirley Hansen. *Managing Indoor Air Quality.* 3d ed. Lilburn, Ga.: Fairmont Press, 2004.

Godish, Thad. *Indoor Environmental Quality.* Boca Raton, Fla.: CRC Press, 2001.

Hines, Anthony L., et al. *Indoor Air: Quality and Control.* Englewood Cliffs, N.J.: Prentice Hall, 1993.

Moffat, Donald W. *Handbook of Indoor Air Quality Management.* Englewood Cliffs, N.J.: Prentice Hall, 1997.

Vallero, Daniel. "Indoor Air Quality." In *Fundamentals of Air Pollution.* 4th ed. Boston: Elsevier, 2008.

# Intergovernmental Panel on Climate Change

CATEGORIES: Organizations and agencies; weather and climate

IDENTIFICATION: International body of scientific experts that evaluates humanity's impact on global climate

DATE: Established in 1988

SIGNIFICANCE: Through the Intergovernmental Panel on Climate Change the international scientific community and the world's governments come together to work cooperatively to compile climate change information for use in formulating policy.

During the 1980's scientists, governments, policy experts, and others began to consider seriously the possibility that human activities might affect the earth's climate in ways that could have broad impacts on vital natural and human-managed systems. In 1988 the World Meteorological Organization (WMO) and the United Nations Environment Programme (UNEP) established the Intergovernmental Panel on Climate Change (IPCC) to conduct an ongoing assessment of scientific and technical information and policy alternatives related to climate change and its possible impacts. Scientists from many nations were invited to participate, with the objective of evaluating the best available scientific research relevant to national and international policy.

The IPCC's membership includes representatives of the world's national governments, who work in cooperation with the organization's elected leaders and the international scientific community. The organization does not conduct original research; rather, it assesses the massive body of climate science literature contained in peer-reviewed scientific journals. So-called gray literature—sources that are unpublished or that have not undergone peer review, such as industry journals, workshop proceedings, internal organizational publications, and reports by governmental agencies and nongovernmental organizations—may also be assessed after the literature's quality and validity have been carefully considered. The IPCC's assessments are intended to inform and support, but not prescribe, climate change policy.

The IPCC has three working groups that carry out its mandate. Working Group I focuses on the science of the earth's climate system. Working Group II assesses scientific, environmental, economic, and social impacts of climate change—both negative and positive—on ecological and socioeconomic systems and human health, emphasizing regional and sectoral analyses. Working Group III examines all aspects of climate change mitigation strategies and response alternatives.

In general, the IPCC's analyses have provided strong support for the theories and projected broad impacts of global warming. The IPCC published its First Assessment Report in 1990, which was followed by a supplementary report in 1992. These publications provided useful summaries of current scientific information in advance of the negotiation of the 1992 United Nations Framework Convention on Climate Change. The IPCC's Second Assessment Report was published in 1995. One sentence from the 530-page Working Group I report has been widely quoted: "The balance of evidence suggests that there is a discernible human influence on global climate." The IPCC released its Third Assessment Report in 2001 and its Fourth Assessment Report in 2007. The Fifth Assessment Report is planned for 2013-2014. The IPCC has also published several supporting documents and technical reports addressing specific scientific and policy issues.

The IPCC's rules of procedure require that each assessment undergo extensive scientific peer review and governmental review processes prior to publication. For the four-volume Fourth Assessment Report, 450 scientists from 130 countries served as lead authors, another 800 were contributing authors, and more than 2,500 experts participated in the peer-review process. An assessment summary for policy makers, which includes an assessment report's key messages, is released only after representatives from participating governments have scrutinized the assessment report thoroughly and approved it by consensus. IPCC authors and reviewers all work on the assessments as unpaid volunteers, as do the panel's chair and elected leaders.

In 2007, the IPCC and Al Gore were jointly awarded the Nobel Peace Prize for their work in building and disseminating knowledge about anthropogenic (human-caused) climate change. Two years later, the organization made headlines for very different reasons. In November, 2009, less than one month before the United Nations Climate Change Conference was to be held in Copenhagen, Denmark, a computer server used by the Climatic Research Unit at the University of East Anglia in the United Kingdom was hacked, and hundreds of e-mail messages, thousands of documents, and source code stolen from the server were posted on various Internet sites. Climate change skeptics cited the contents of some of the correspondence between IPCC contributing researchers as evidence of deliberate manipulation and suppression of data. The scientists who were involved countered that the correspondence excerpts were taken out of context or were deliberately misinterpreted in the worst possible light, and that they revealed no unethical behavior (although some angrily worded e-mails reflected the frustrations of working in such a politically contentious field). The IPCC stood behind its scientific community and defended the Fourth Assessment Report as being comprehensive, unbiased, open, and transparent. The

IPCC emphasized that its thoroughgoing assessment procedures made it impossible for any individual or small group to omit or distort findings or change conclusions. In March and July, 2010, three official inquiries all found that the scientists at the Climatic Research Unit had not manipulated any data.

The IPCC's credibility suffered another blow in January, 2010, after news stories reported that Working Group II's contribution to the 2007 assessment report had referenced poorly substantiated and overly pessimistic estimates regarding the fate of Himalaya's glaciers—information from a gray-literature source. The IPCC expressed regret over the improper application of its well-established standards of evidence and reaffirmed its commitment to those standards.

*Phillip A. Greenberg*
*Updated by Karen N. Kähler*

FURTHER READING

Bolin, Bert. *A History of the Science and Politics of Climate Change: The Role of the Intergovernmental Panel on Climate Change.* New York: Cambridge University Press, 2007.

Dessler, Andrew Emory, and Edward Parson. *The Science and Politics of Global Climate Change: A Guide to the Debate.* New York: Cambridge University Press, 2006.

Intergovernmental Panel on Climate Change. *Climate Change 2007: Synthesis Report.* Geneva: Author, 2008.

# Krakatoa eruption

CATEGORIES: Disasters; atmosphere and air pollution
THE EVENT: Massive volcanic eruption that took place in the Malay Archipelago
DATES: August 26-27, 1883
SIGNIFICANCE: The eruption of the volcano Krakatoa ejected tons of rock, ash, and gases into the atmosphere, cooling the planet and causing sea levels to fall.

The volcano Krakatoa (or Krakatau) erupted in a series of four spectacular explosions on August 26 and 27, 1883. The eruption was the fourth largest in recorded history and destroyed two-thirds of the island— also known as Krakatoa—on which the volcano stood. The cataclysm resulted in more than 36,000 deaths,

Krakatoa in Modern Indonesia

most of them on the nearby islands of Java and Sumatra, and had profound worldwide consequences as well.

Krakatoa's eruption produced a shock wave that passed around the earth seven times, and its sound was clearly heard 4,777 kilometers (2,968 miles) away. The tsunamis that resulted from the eruption, which caused many deaths in the immediate vicinity, were detected as ripples as far away as the English Channel. An estimated 21 to 25 cubic kilometers (5 to 6 cubic miles) of material was ejected into the atmosphere by the volcano, some of it as high as 48 kilometers (30 miles). Ash drifted down on ships thousands of miles away, but much of the finer material remained floating in the stratosphere and was carried around the earth. Vivid atmospheric effects of the eruption, including red and yellow sunsets and halos around the sun and moon, were reported from widely scattered locations for three years. Such observations provided scientists with their first proof of a worldwide system of high-altitude winds.

Of greater importance to the larger environment, sulfur dioxide ($SO_2$) gas emitted during the eruption combined with water vapor to create droplets of sulfuric acid ($H_2SO_4$) in the upper atmosphere. Together with floating ash, these droplets reduced the sunlight striking the earth by about 1 percent for two years, cooling the planet as much as 0.5 degree Celsius (0.9 degree Fahrenheit) for approximately five years. In 2006 P. J. Gleckler and five colleagues determined that the reduction in sunlight also caused the oceans

to cool and, as a result, sea levels to fall. Sea levels regained their previous height only during the middle of the twentieth century.

The fragments of Krakatoa remaining after the eruption were covered by as much as 40 meters (131 feet) of ash, and it was assumed that every living organism on the island had died. For scientists concerned with the recolonization of such a landscape by plants and animals, the site became a living laboratory. A few months after the eruption a small spider was observed, and six years later a monitor lizard and a variety of insects were noted. Within a few decades, trees had sprouted and grown to more than 12 meters (39 feet) in height.

In late June, 1927, fishermen noticed steam rising from the sea above the collapsed portions of Krakatoa, indicating a renewal of volcanic activity. On January 26, 1928, a new island appeared above the surface of the sea and was soon named Anak (child of) Krakatoa. It has since grown steadily and has erupted several times. The Krakatoa area was declared a nature reserve in 1921, and it was later made a part of Indonesia's Ujung Kulon National Park. In 1991 the park was designated a World Heritage Site by the United Nations Educational, Scientific, and Cultural Organization (UNESCO).

*Grove Koger*

FURTHER READING

Gleckler, P. J., et al. "Krakatoa's Signature Persists in the Ocean." *Nature* 439 (February 9, 2006): 675.

Simkin, Tom, and Richard S. Fiske. *Krakatau, 1883: The Volcanic Eruption and Its Effects.* Washington, D.C.: Smithsonian Institution Press, 1983.

Winchester, Simon. *Krakatoa: The Day the World Exploded, August 27, 1883.* New York: HarperCollins, 2003.

# Kyoto Protocol

CATEGORIES: Treaties, laws, and court cases; weather and climate

THE TREATY: International agreement that commits nations to place legally binding limits on their emissions of six greenhouse gases

DATE: Adopted on December 11, 1997

SIGNIFICANCE: The Kyoto Protocol represents the first international, legally binding attempt to prevent human activity from causing significant adverse changes to the earth's climate.

In 1988 the World Meteorological Organization and the United Nations established the Intergovernmental Panel on Climate Change (IPCC), a team of more than two thousand leading scientists from around the world whose mission would be to assess scientific information on climate and the environmental impacts of climate change. In 1990 the IPCC released its first report, which concluded that human-made greenhouse gases would exacerbate the greenhouse effect, resulting in additional warming of the earth's surface by the twenty-first century unless measures were enacted to limit the emissions of these gases.

At the 1992 Earth Summit in Brazil, the United Nations Framework Convention on Climate Change (UNFCCC) was adopted. This treaty, signed by more than 150 nations, required each nation to limit its greenhouse gas emissions, with the industrialized nations taking the first step by voluntarily reducing their emissions to 1990 levels by the year 2000. U.S. participation in the treaty was ratified by the U.S. Senate in October, 1992, and eventually more than 160 nations joined in the agreement.

In 1995 the IPCC released a report stating for the first time that the balance of evidence suggested a "discernible human influence on global climate." The report noted that even if emission levels were to remain constant, the atmospheric concentrations of carbon dioxide would approach twice the prein-

dustrial concentration by the end of the twenty-first century, changing the earth's climate in significant ways. Between 1992 and 1995 global greenhouse emissions continued to rise, and it was agreed that the voluntary approach had not been successful. At a meeting in Berlin, Germany, in 1995, negotiators agreed that the industrialized nations, which emit the majority of greenhouse gases, would have to take the lead in adopting stronger measures. In 1996 the United States announced that future emission targets should be legally binding and challenged other industrialized nations to agree. More than one hundred nations agreed to develop legally binding targets.

In a March, 1997, meeting in Bonn, Germany, the European Union took the lead by proposing that industrialized nations reduce emissions by 15 percent from 1990 levels by the year 2010. The U.S. government proposed a system of international trading of emissions rights, in which nations could buy and sell the rights to emit greenhouse gases. The United States also proposed a "joint implementation" program that would allow nations to earn emissions credits by implementing non-carbon-based "clean energy" projects. At the same meeting, the IPCC chairperson reported that reductions undertaken solely by industrialized nations would not be sufficient to limit global warming to environmentally sustainable levels.

As the December, 1997, Conference to the Parties to the UNFCCC in Kyoto, Japan, drew near, it appeared that reaching a consensus would be difficult, as proposed policies ranged from a 20 percent reduction (compared to 1990 levels) by 2005 to the U.S. proposal of merely stabilizing emissions at 1990 levels. Most developing nations stated that they would not commit to emissions controls until after the developed nations had acted.

As talks opened, the United States made a significant change in its position by announcing that it would support a system of flexible targets for different nations that would take into consideration the situation of each country. The United States maintained its position of zero reduction compared to 1990 levels until late in the conference, when Vice President Al Gore instructed the U.S. delegation to be more flexible in negotiations. Gore's involvement helped to break the logjam of deliberations, and the final days of the conference were marked by around-the-clock negotiations until an agreement was finalized during the last hours.

The Kyoto Protocol called for a 5.2 percent reduction in emissions of carbon dioxide, methane, and nitrous oxide from 1990 levels by the period from 2008 to 2012, with the United States reducing emissions by 7 percent, Japan by 6 percent, and the European Union by 8 percent. Three other greenhouse gases, all chlorofluorocarbon (CFC) substitutes, were to be cut by comparable levels, using 1995 rather than 1990 as the baseline. While the accords reached in Kyoto marked the beginning of a legal framework for reducing carbon emissions over the long run, the emission cuts required by the treaty fell far short of the levels that most climate scientists predicted would be needed to prevent significant climate change.

The Kyoto Protocol was criticized by many environmental leaders as having major loopholes, such as "flexibility measures" that would allow nations to circumvent their requirements for reductions in emissions. Rules for how industrialized nations could trade or sell emission rights were not formalized in the agreement. Russia was left with emission credits because its emissions plummeted during the collapse of its economy in the 1990's—these credits could be sold to countries such as the United States, with the result that there might be little or no reduction in emissions by major polluters. Also not addressed in the agreement was how joint implementation would affect emissions reduction targets. Finally, no compliance mechanisms were included.

The most contentious issue resulting from the Kyoto conference concerned the exclusion of developing countries from the accords. At the conference, representatives from a united group of 130 developing nations, led by China and India, voiced strong objections to the process of emissions trading, claiming that such trades would allow rich polluters to buy their way out of reductions. At the conference, the United States demanded voluntary commitments from developing nations to reduce emissions, but the proposal was rejected in the final negotiations. On a per capita basis, emissions in developing nations were well below those of the industrialized nations. However, emissions were growing so rapidly in developing nations that they were projected to exceed those of the industrialized nations by 2025, with China projected to be the largest emitter of carbon dioxide by 2015. Many industrialized nations, including the United States, refused to ratify the treaty unless the developing nations were included.

The Kyoto Protocol represented the first international legally binding attempt to prevent human activity from causing significant adverse changes to the

## Emissions Trading

*Article 6 of the Kyoto Protocol, reproduced below, establishes the basic framework for parties to the treaty to trade pollution credits with one another, thereby employing market principles to drive international emission reductions.*

1. For the purpose of meeting its commitments under Article 3, any Party included in Annex I may transfer to, or acquire from, any other such Party emission reduction units resulting from projects aimed at reducing anthropogenic emissions by sources or enhancing anthropogenic removals by sinks of greenhouse gases in any sector of the economy, provided that:

(a) Any such project has the approval of the Parties involved;

(b) Any such project provides a reduction in emissions by sources, or an enhancement of removals by sinks, that is additional to any that would otherwise occur;

(c) It does not acquire any emission reduction units if it is not in compliance with its obligations under Articles 5 and 7; and

(d) The acquisition of emission reduction units shall be supplemental to domestic actions for the purposes of meeting commitments under Article 3.

2. The Conference of the Parties serving as the meeting of the Parties to this Protocol may, at its first session or as soon as practicable thereafter, further elaborate guidelines for the implementation of this Article, including for verification and reporting.

3. A Party included in Annex I may authorize legal entities to participate, under its responsibility, in actions leading to the generation, transfer or acquisition under this Article of emission reduction units.

4. If a question of implementation by a Party included in Annex I of the requirements referred to in this Article is identified in accordance with the relevant provisions of Article 8, transfers and acquisitions of emission reduction units may continue to be made after the question has been identified, provided that any such units may not be used by a Party to meet its commitments under Article 3 until any issue of compliance is resolved.

earth's climate. For the treaty to enter into force, however, it had to be ratified by governments representing at least 55 nations, including industrialized nations representing 55 percent of all 1990 carbon dioxide emissions. Successful implementation of the protocol thus awaited the resolution of issues of nationalism, ideology, and economics. Continued discussions among nations resulted in steady progress on these issues, and the requirement of 55 signatory nations was met in 2002. When Russia signed in 2004, the requirement that signatories include industrialized nations representing 55 percent of 1990 carbon dioxide emissions was satisfied, allowing the treaty to enter into force soon after, on February 16, 2005. By 2009 the treaty had been ratified by 186 nations and the European Union.

*Craig S. Gilman*

FURTHER READING

Conkin, Paul K. "Greenhouse Gases and Climate Change." In *The State of the Earth: Environmental Challenges on the Road to 2100*. Lexington: University Press of Kentucky, 2007.

Dessler, Andrew E., and Edward A. Parson. *The Science and Politics of Global Climate Change: A Guide to the Debate*. New York: Cambridge University Press, 2006.

Press, Frank, et al. "Earth's Environment, Global Change, and Human Impacts." In *Understanding Earth*. 4th ed. New York: W. H. Freeman, 2004.

Zedillo, Ernesto, ed. *Global Warming: Looking Beyond Kyoto*. Washington, D.C.: Brookings Institution Press, 2008.

# London smog disaster

CATEGORIES: Disasters; atmosphere and air pollution

THE EVENT: Incident in which a lethal combination of fog, smoke, and pollutants settled over London, England, for several days

DATES: December 4-8, 1952

SIGNIFICANCE: After London's "killer smog" incident resulted in thousands of illnesses and deaths, the British government undertook to pass strong legislation that would address the problem of air pollution.

On the evening of December 4, 1952, weather conditions caused a "killer smog"—a lethal combination of fog, smoke, and pollutants—to settle over London, England; the heavy smog did not clear until December 9. Transportation was severely disrupted by low visibility during the disaster, and many outdoor sporting events had to be canceled. The chief culprit in the London smog disaster was deemed to be Great Britain's heavy dependence on coal, which was used by industry and burned in almost every household in London, since the country's wood supplies had long been depleted. Particular blame was placed on the popular "nutty slack," a low-grade soft coal used in most households that burned inefficiently and gave off noxious smoke and odor. Authorities estimated that approximately one-half of the smoke emitted during the killer smog came from household hearths.

While the smog was present and especially in the period immediately after the smog abated, the health effects of breathing such polluted air for days began to be seen. Thousands were hospitalized with circulatory and respiratory problems, including bronchitis, influenza, and pneumonia. During the last three weeks of December, the death rate in London climbed dramatically. The minister of health later announced that during a five-week period ending January 3, 1953, more than fifteen thousand deaths were registered in Greater London, compared with approximately nine thousand during the same period one year previously. It has been estimated that at least four thousand deaths were directly attributable to the smog, with most victims being babies under one year old and persons over the age of fifty-five. The death rate for this period exceeded that of previous disasters in London's history, including the worst periods of the cholera epidemic of 1866 and the great fog of December, 1873.

The British Clean Air Act of 1956, which represented an attempt to correct the smog problem, was a direct result of the 1952 disaster. The act banned the emission of black smoke from locomotives, vessels, and chimneys, and it required that all new furnaces be capable, as far as practical, of not producing smoke. In addition, the act banned the emission of grit or dust from all furnaces, new or old. It also created the Clean Air Council to advise government ministers in regard to clean air policy. One of the act's most important provisions gave cities the power to establish "smoke-control zones" to combat the problems caused by the burning of coal in home fireplaces and at local factories. To assist in this last objective, the national government offered financial grants to cover 40 percent of the cost of converting home appliances so that they could burn smokeless fuel; local government authorities were required to contribute another 30 percent, so that the appliance owners had to pay only 30 percent of the total cost of conversion.

Even with the measures initiated under the Clean Air Act, it was estimated that it would take up to fifteen years for the full impact of the improvements to be felt. London's smog problem was eventually overcome, however, thanks to this legislation and to the long-term trend of factories, railroads, and households converting to oil, gas, nuclear, and electrical energy.

*David C. Lukowitz*

## FURTHER READING

Benton-Short, Lisa, and John R. Short. *Cities and Nature.* New York: Routledge, 2008.

Kessel, Anthony. *Air, the Environment, and Public Health.* New York: Cambridge University Press, 2006.

Vallero, Daniel. *Fundamentals of Air Pollution.* 4th ed. Boston: Elsevier, 2008.

# Malathion

CATEGORY: Pollutants and toxins

DEFINITION: An organophosphorus pesticide used against mosquitoes, fleas, and other insects

SIGNIFICANCE: Although no adverse health effects have been found from low-level exposure to malathion, the use of this pesticide in aerial spraying programs has been controversial.

Pesticides have been used to control crop loss and prevent the spread of disease for hundreds of years. Until the middle of the twentieth century, pesticides were either toxic metals (such as arsenic or lead) or natural products (such as nicotine sulfate). With the discovery of dichloro-diphenyl-trichloroethane (DDT) in 1939, new human-made pesticides became available for use. However, it was soon found that DDT and related chlorinated hydrocarbon pesticides persist in the environment for years, causing damage to birds, fish, and other wildlife. Other pesticides, including organophosphate compounds, were therefore developed as alternatives to DDT.

Malathion is an organophosphate compound that is chemically similar to some types of nerve gas but less toxic than most other organophosphate pesticides. Unlike chlorinated hydrocarbons, malathion breaks down in the environment, transforming into carbonic and phosphoric acid over a period of days to a few weeks. Because of its relatively short lifetime, malathion does not bioaccumulate in aquatic organisms or contaminate groundwater. Following its introduction by the American Cyanamid Company in 1952, malathion quickly became a popular substitute for DDT in controlling mosquitoes, fleas, lice, and other insects. In 1956 malathion was first used to counteract fruit fly infestations.

In the following decades, the use of malathion continued to increase. In the 1980's, however, the spraying of malathion in California to control the Mediterranean fruit fly (also known as the Medfly) became a source of controversy. Those critical of the use of malathion in populated areas noted that commercial preparations of malathion contain trace impurities that are potentially toxic and that the initial breakdown products from malathion include malaoxin, a compound with acute toxicity forty times that of malathion itself. Critics also suggested that malathion might be carcinogenic or could weaken the immune system, making people more susceptible to disease. However, extensive studies of human populations exposed to malathion through the spraying program found no adverse health effects from low levels of exposure. Similarly, no harmful effects have been observed from low-level exposure to malathion in laboratory studies on animals. Despite the controversy, malathion continued to be used to prevent damage to citrus crops in California, Florida, and other states, and to be available in many commercial products.

By the end of the twentieth century, controversy over the use of synthetic pesticides such as malathion had convinced many people that the release of such compounds into the environment should be greatly reduced, but malathion continued to be used in some areas into the twenty-first century: Both New York City and the city of Winnipeg in Manitoba, Canada, used malathion spraying (in 2000 and 2005, respectively) as part of their attempts to eradicate insects, particularly mosquitoes, that might be carriers of West Nile virus. Critics of human-made pesticides have continued to call for more intelligent use of pesticides and the development of alternative methods of controlling insect populations.

*Jeffrey A. Joens*

FURTHER READING

Hamilton, Denis, and Stephen Crossley, eds. *Pesticide Residues in Food and Drinking Water: Human Exposure and Risks.* Hoboken, N.J.: John Wiley & Sons, 2004.

Levine, Marvin J. "Pesticides in Food." In *Pesticides: A Toxic Time Bomb in Our Midst.* Westport, Conn.: Praeger, 2007.

Manahan, Stanley E. *Fundamentals of Environmental Chemistry.* 3d ed. Boca Raton, Fla.: CRC Press, 2009.

## *Massachusetts v. Environmental Protection Agency*

CATEGORIES: Treaties, laws, and court cases; weather and climate

THE CASE: U.S. Supreme Court ruling concerning the regulation of greenhouse gases

DATE: Decided on April 2, 2007

SIGNIFICANCE: The Supreme Court's decision in the case of *Massachusetts v. Environmental Protection Agency* substantially enhanced the opportunity of states to challenge the decisions of federal agencies in court, but it did not quickly produce an EPA regulation governing greenhouse gas emissions.

In 1999 a group of environmental organizations petitioned the U.S. Environmental Protection Agency (EPA) to exercise its authority under the Clean Air Act to make a rule regulating greenhouse gases from vehicle emissions. In 2001 the EPA requested public comment, and in 2003 the agency denied the petition, giving three reasons: First, the EPA lacked the authority under the Clean Air Act to issue carbon dioxide emission standards; second, scientific evidence had not conclusively established causal links between human activity, an increase in greenhouse gases, and a rise in global air temperature; and third, EPA regulation of greenhouse gases would conflict with the president's climate change policy and hamper his ability to negotiate with other nations to reduce greenhouse gases. Massachusetts, joined by state and local governments and environmental organizations, appealed the decision. After the U.S. Court of Appeals denied review in 2005, the state appealed to the U.S. Supreme Court.

The Supreme Court first addressed the issue of the state's standing to invoke the Court's jurisdiction. Justice John P. Stevens, writing for the Court, stated that the Court relaxed its standing test for two reasons: First, the dispute involved the interpretation of a federal statute, the Clean Air Act, intended to protect Massachusetts; and second, the party seeking review was a sovereign state that owned the territory alleged to be harmed, an interest that was entitled to particular generosity. Applying its three-part standing test, the Court granted Massachusetts standing to challenge the EPA's refusal to grant the state's rule-making petition, because the state had demonstrated sufficient injury by alleging that it had lost a significant portion of its coastline, that the EPA's refusal to regulate auto emissions (though such emissions are only one source of greenhouse gases) contributed to this injury, and that the EPA's use of its Clean Air Act authority to regulate carbon dioxide from new motor vehicles would be likely to redress the injury by slowing global warming.

The Court then addressed the two issues that had been appealed. First, it held that the EPA had authority under the Clean Air Act to regulate greenhouse gas emissions from new motor vehicles because the statute's broad definition of "air pollutant" includes all airborne pollutants. Second, it held that the EPA's exercise of its authority had to be grounded in the Clean Air Act, which granted the agency discretion to determine if an air pollutant causes or contributes to the endangerment of public health or welfare, not the discretion to make policy arguments that it would be unwise to regulate at the present time. Since the EPA provided no reasoned explanation for its denial of the rule-making petition, its action violated the Clean Air Act. The EPA was mandated to provide a reasoned explanation for its decision grounded in the statute.

The Supreme Court's decision substantially enhanced the opportunity of states to challenge federal agency decisions in court, but it did not quickly produce an EPA regulation governing greenhouse gas emissions. In May, 2007, President George W. Bush directed the EPA to write new regulations, but by the end of 2008, none had been issued. In April, 2009, the EPA under President Barack Obama's administration concluded its scientific review and announced a proposed finding that greenhouse gases, including carbon dioxide, endanger public health and welfare. Then, on December 7, 2009, the opening day of the United Nations Climate Change Conference in Copenhagen, Denmark, the EPA finalized its endangerment finding. This decision began the agency's process of writing a new rule regarding the emissions from motor vehicles, power plants, and factories that contribute to global warming.

*William Crawford Green*

FURTHER READING

Ferrey, Steven. "Air Quality Regulation." In *Environmental Law: Examples and Explanations*. 5th ed. New York: Aspen, 2010.

Heinzerling, Lisa. "The Role of Science in *Massachusetts v. EPA*." *Emory Law Journal* 58 (2008): 411-422.

Sugar, Michael. "*Massachusetts v. Environmental Protection Agency*." *Harvard Environmental Law Review* 31 (2007): 531-544.

# Medfly spraying

CATEGORY: Pollutants and toxins

DEFINITION: Aerial application of pesticides to eliminate the Mediterranean fruit fly, an agricultural threat

SIGNIFICANCE: Efforts to eradicate Medfly infestations through the spraying of pesticides have met with criticism by many environmentalists because of concerns about the possible harmful effects of the pesticides on other insects, wildlife, livestock, aquaculture, water supplies, and human health.

The small, two-winged Mediterranean fruit fly, or Medfly (*Ceratitis capitata*), belongs to a group of insects commonly called fruit flies. Considered a major agricultural pest around the world, the Medfly is a threat to more than 250 vegetables and fruits, including peaches, cherries, avocados, pears, and citrus fruits. A longtime inhabitant of Hawaii and the Tropics, the Medfly has repeatedly tried to establish itself in the continental United States since its first unsuccessful foray into Florida in 1929. Attempts to control the fly with pesticides have so far proven successful but controversial.

The incompatibility of the Medfly with humans stems from the insect's fondness for domestic crops. Typically, the female fly lays eggs—as many as several hundred at a time—in the fleshy parts of fruits or vegetables. When the eggs hatch, the larvae tunnel through the fruit, making it unfit for human consumption. When the damaged fruit falls to the ground, the larvae exit and burrow into the ground until they mature into flies and start the cycle over again.

Because of the Medfly's capacity for causing widespread crop destruction, governments around the world have imposed various quarantines, embargoes, and postharvest treatment requirements on any fruits and vegetables that originate in areas known to be infested with the insect. Various eradication programs were used against the Medfly in the United States during the twentieth century. Early efforts in Florida included the use of a compound of arsenate and copper carbonate that was applied with handheld equipment. Another control method was the removal of infested fruit trees.

The most common form of eradication practiced is the aerial spraying of the pesticide malathion mixed with a bait, such as syrup. A poison also employed to control mosquitoes, malathion is used in weaker amounts in the war against the Medfly. Although all aerial spraying attempts in the United States must receive prior approval from the Environmental Protection Agency (EPA), the use of malathion has generated controversy in California and Florida. Opponents commonly complain about the pesticide's possible harmful effects on other insects, wildlife, livestock, aquaculture, water supplies, and human health—especially that of children. In response to these concerns, the EPA now requires various federal and state agencies involved in eradication programs to seek more environmentally friendly methods to battle the Medfly.

One alternative to the spraying of malathion involves the use of domestically raised flies that have been sterilized through radiation; these flies are released to mate unsuccessfully with wild Medflies, thus reducing the regenerative capacity of the population. Another method that has been used involves a pesticide made from a mixture of two dyes, phloxine B and uranine, which are commonly used to tint drugs and cosmetics. Once ingested by the Medfly, the dye particles absorb light, which in turn produces oxidizing agents that destroy cell tissues. As a result, most flies die within twenty-four hours. The dyes quickly lose their potency and become nontoxic. Preliminary tests of this method in Hawaii indicated that the use of these dyes may be more effective and safer for the environment than malathion.

*John M. Dunn*

FURTHER READING

Gullan, P. J., and P. S. Cranston. "Pest Management."
In *The Insects: An Outline of Entomology.* 4th ed.
Hoboken, N.J.: John Wiley & Sons, 2010.

Hamilton, Denis, and Stephen Crossley, eds. *Pesticide
Residues in Food and Drinking Water: Human Exposure
and Risks.* Hoboken, N.J.: John Wiley & Sons, 2004.

# Methane

CATEGORIES: Pollutants and toxins; atmosphere and
air pollution; weather and climate

DEFINITION: Colorless, odorless gas that is the principal component of natural gas

SIGNIFICANCE: Methane in the form of natural gas represents an abundant source of energy, and methane
combustion is cleaner than petroleum or coal combustion. Methane is a powerful greenhouse gas,
however; scientists are thus faced with the challenge
of optimizing energy production using methane
while minimizing unwanted methane emissions.

Methane is produced when bacteria di-
gest organic matter under anaerobic
(without air) conditions, creating natural
gas. Natural gas contains 50-90 percent
methane. Most natural gas is found with
coal and petroleum deposits buried deep
underground and is a product of the decomposition of ancient swamps and bogs.

The sources of methane emissions are
both anthropogenic (human-influenced)
and natural. Anthropogenic sources include fossil-fuel production, livestock raising (enteric fermentation in the stomachs
of animals such as cattle and pigs produces
approximately 37 percent of all human-induced methane), rice cultivation, biomass
burning, and waste management (sewage
treatment and landfills). The Intergovernmental Panel on Climate Change estimates that more than 60 percent of global
methane emissions are related to such activities. Natural sources of methane emissions include wetlands, which provide habitat conducive to bacteria that produce
methane during their decomposition of
organic material; the digestive processes

of termites (the second-largest natural source of
methane emissions); and oceans, where methane
emissions come from anaerobic digestion by marine
zooplankton and fish and by methanogenesis in marine sediments.

Methane is also stored as a hydrate (methane hydrate, a crystalline solid consisting of gas molecules
surrounded by a cage of water molecules) in immense
amounts in marine sediments and in the frozen Arctic
tundra. The worldwide amount of carbon stored as
methane hydrate is estimated to total twice the
amount of carbon found in all known fossil fuels on
earth. Methane can be released from the hydrates by
increases in temperature and other factors.

Methane is used in industrial and chemical processes and can be transported as a refrigerated liquid
(liquefied natural gas, or LNG). It is useful as a fuel for
cooking, for powering motor vehicles, and for heating homes and commercial buildings. Many factory
furnaces burn methane, and it is also used to generate
electricity. Compressed natural gas is thought to be
the cleanest-burning form of fossil fuel available,
since the simplicity of the methane molecule results

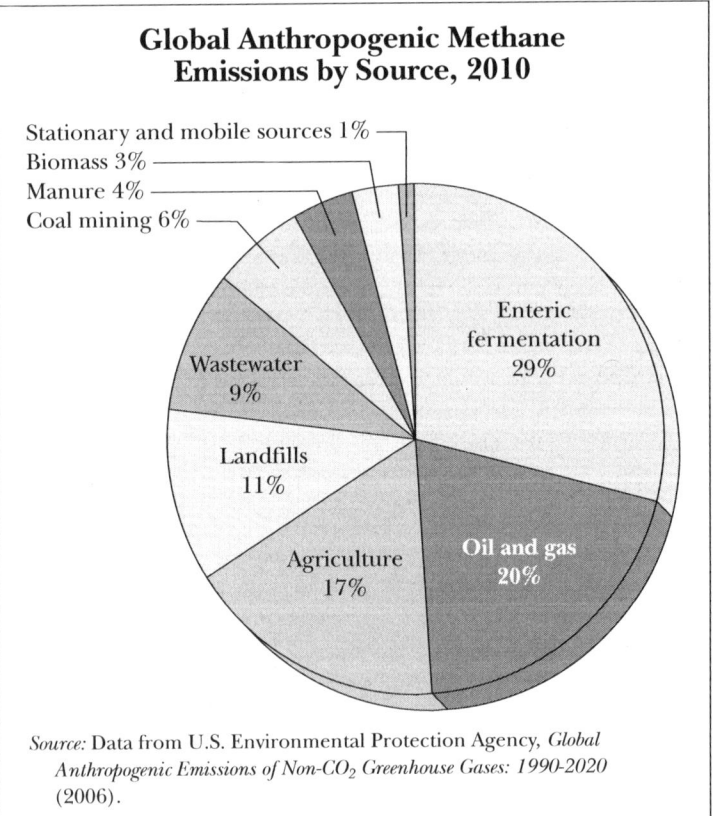

**Global Anthropogenic Methane Emissions by Source, 2010**

Stationary and mobile sources 1%
Biomass 3%
Manure 4%
Coal mining 6%
Wastewater 9%
Landfills 11%
Agriculture 17%
Enteric fermentation 29%
Oil and gas 20%

*Source:* Data from U.S. Environmental Protection Agency, *Global
Anthropogenic Emissions of Non-CO$_2$ Greenhouse Gases: 1990-2020*
(2006).

in lower emissions of various pollutants than are produced by other fossil fuels. Methane is also a key raw material for producing solvents (such as methanol) and other organic chemicals.

## METHANE REMOVAL FROM THE ATMOSPHERE

According to the U.S. Environmental Protection Agency (EPA), released methane can remain in the earth's atmosphere for nine to fifteen years. Once emitted, methane can be removed from the atmosphere by a variety of processes, frequently called sinks. The dominant sink is oxidation by photochemically produced hydroxyl radicals (OH). The majority of methane molecules react with OH to form the methyl group $CH_3$ and water in the tropospheric layer of the atmosphere, with smaller amounts of methane destroyed in the stratosphere. These two OH reactions account for almost 90 percent of methane removal. Two smaller sinks are microbial uptake of methane in soils and methane's reaction with chlorine atoms in the marine boundary layer.

The balance between methane emissions and methane removal processes ultimately determines atmospheric methane concentrations. The methane remaining after methane removal processes can absorb terrestrial infrared radiation that would otherwise escape to space. This property can contribute to the warming of the atmosphere, which is why methane is considered to be a greenhouse gas. It is more than twenty times more effective in trapping heat in the atmosphere than carbon dioxide ($CO_2$) over a period of one hundred years.

## RESEARCH DIRECTIONS

Methane research is proceeding in two major directions: Some focuses on energy generation, seeking ways to make bioconversion of wastes to methane more economically attractive as an alternative fuel, and some focuses on the environment, seeking ways to limit the release of methane into the atmosphere because of its properties as a greenhouse gas. A shared goal of these two areas of research is to find ways in which much of the methane released into the atmosphere could be harnessed for energy production. Methane has advantages over petroleum and coal as a fuel because it burns more cleanly than do those fossil fuels.

In the energy sector, operators of coal mines are looking for ways to isolate the methane produced as a result of mining activities, instead of venting it into the atmosphere. The EPA estimates that up to 40 percent of the methane that migrates to the atmosphere can be used for heat and power generation, injection into pipeline systems for transport, methanol production, or on-site applications such as coal drying. Landfill gas-to-energy projects, which collect the methane that forms in landfills, offer another promising way to reduce atmospheric release and provide inexpensive energy.

The U.S. Global Change Research Program has identified as a priority research activity the development of global monitoring sites to measure atmospheric methane levels. The Carbon Cycle Greenhouse Gases Group of the National Oceanic and Atmospheric Administration's Climate Monitoring and Diagnostics Laboratory also makes ongoing atmospheric methane measurements from land and sea surface sites and aircraft and continuous measurements from baseline observatories and towers. Measurement records from international laboratories are integrated and extended to produce a globally consistent cooperative data product called GLOBALVIEW.

*Bernard Jacobson*

## FURTHER READING

Buell, Phyllis, and James Girard. *Chemistry Fundamentals: An Environmental Perspective.* 2d ed. Sudbury, Mass.: Jones and Bartlett, 2003.

Demirbas, Ayhan. *Methane Gas Hydrate.* London: Springer, 2010.

Khalil, M. A. K., ed. *Atmospheric Methane: Its Role in the Global Environment.* New York: Springer, 2000.

National Academy of Sciences. *Climate Change Science: An Analysis of Some Key Questions.* Washington, D.C.: National Academies Press, 2001.

_____. *Methane Generation from Human, Animal, and Agricultural Wastes.* Washington, D.C.: National Academies Press, 2001.

Soliva, Carla Riccarda, Junichi Takahashi, and Michael Kreuzer, eds. *Greenhouse Gases and Animal Agriculture.* Boston: Elsevier, 2006.

# Molina, Mario

CATEGORY: Weather and climate
IDENTIFICATION: Mexican chemist
BORN: March 19, 1943; Mexico City, Mexico
SIGNIFICANCE: Molina's pioneering work concerning
the formation and catalytic decomposition of
ozone in the stratosphere led to greater scientific
attention to the issue of climate change.

The ozone that is produced naturally in the earth's
stratosphere provides a filter to restrict the
amount of harmful ultraviolet radiation that strikes
the surface of the planet. It was the work of Mario
Molina, Paul Crutzen, and Frank Sherwood Rowland
that first drew attention to the catalytic role played by
chlorofluorocarbons (CFCs) in the decomposition of
ozone. CFCs were long used in aerosol spray cans,
electronic parts cleaning, and refrigeration systems.
Their production was banned in 1987 by an international agreement called the Montreal Protocol.

Molina was born in Mexico City in 1943 to Roberto
Molina Pasquel and Leonor Henríquez de Molina.
He obtained a degree in chemical engineering in
1965 from the Universidad Nacional Autónoma de
México. He earned a master's degree at the University
of Freiburg in Germany in 1967 for work done on the
study of rates of polymerization reactions. Molina returned to his undergraduate institution as an assistant
professor for one year and then began working on a
doctorate at the Berkeley campus of the University of
California. He earned his Ph.D. in 1972 in physical
chemistry and remained at Berkeley as a postdoctoral
student.

In 1973 Molina accepted a postdoctoral appoint-
ment with Rowland at the University of California at
Irvine. At this time Molina began to look at the question of what happens to chemicals and pollutants in
the troposphere and stratosphere, where they are subject to high-intensity ultraviolet radiation. A 1974 article by Molina in the journal *Nature* began a series of
papers pointing out the connection between ozone
destruction and CFCs. Molina's continuing research
led to the prediction of the ozone hole that was later
discovered over Antarctica in 1985. Molina then began researching the interface between the atmosphere and the biosphere; he hoped his investigations
would lead to an understanding of global climate
change processes.

Molina has held appointments with the Jet Propul-
sion Laboratory at the California Institute of Technology as a senior research scientist and with the Department of Earth, Atmospheric, and Planetary Sciences
and the Department of Chemistry at the Massachusetts Institute of Technology (MIT) as a professor. In
2004, he accepted positions in the Department of
Chemistry and Biochemistry at the University of California, San Diego, and at the Center for Atmospheric
Sciences at the Scripps Institution of Oceanography.

Molina's work has been recognized with numerous
honors, including the American Chemical Society
Esselen Award (1987), the American Association for
the Advancement of Science Newcomb-Cleveland
Prize (1988), the National Aeronautics and Space Administration Medal for Exceptional Scientific Advancement (1989), and the United Nations Environment Programme's Global 500 Award (1989). In
1995, Molina shared the Nobel Prize in Chemistry
with Crutzen and Rowland for his pioneering work
concerning the formation and catalytic decomposition of ozone in the stratosphere.

*Kenneth H. Brown*

FURTHER READING
Hill, Marquita K. "Stratospheric Ozone Depletion."
In *Understanding Environmental Pollution.* 3d ed.
New York: Cambridge University Press, 2010.
Joesten, Melvin D., John L. Hogg, and Mary E.
Castellion. "Chlorofluorocarbons and the Ozone
Layer." In *The World of Chemistry: Essentials.* 4th ed.
Belmont, Calif.: Thomson Brooks/Cole, 2007.
Miller, G. Tyler, Jr., and Scott Spoolman. "Climate
Change and Ozone Depletion." In *Environmental
Science: Problems, Concepts, and Solutions.* 13th ed.
Belmont, Calif.: Brooks/Cole, 2010.

# Montreal Protocol

CATEGORIES: Treaties, laws, and court cases;
atmosphere and air pollution
THE TREATY: International agreement with the specific goal of limiting atmospheric inputs of chlorofluorocarbons for the purpose of protecting the
ozone layer
DATE: Opened for signature on September 16, 1987
SIGNIFICANCE: The Montreal Protocol was a political
innovation in that it called for a gradual reduction

in chlorofluorocarbon production and was created to be flexible enough to respond to new scientific information. The agreement led to reductions in atmospheric concentrations of chemical compounds known to deplete ozone in the earth's atmosphere.

The ozone layer, which is 10 to 20 kilometers (6 to 12 miles) above the earth's surface, screens out most of the sun's ultraviolet radiation. Ultraviolet radiation can lead to mutations and cancer in living things. The nations participating in the Montreal Protocol were motivated to act by four major developments: The accumulation of chlorofluorocarbons (CFCs) in the atmosphere was observed in the early 1970's; CFC decomposition in the atmosphere was demonstrated to cause ozone destruction in 1974; a hole in the ozone layer was discovered over Antarctica in the early 1980's, and scientists found evidence linking the ozone hole to CFCs in 1985; and CFC substitutes were developed by important CFC producers.

On September 16, 1987, the Montreal Protocol on Substances That Deplete the Ozone Layer was signed by 46 nations, including the United States. These nations represented two-thirds of the total global production and consumption of CFCs and other halogenated compounds. The protocol entered into force on January 1, 1989. By 2010, 196 nations had ratified the treaty.

The Montreal Protocol was designed to control the production and consumption of CFCs and other halogenated compounds suspected of causing ozone destruction. The United States and other industrialized countries committed themselves to freezing CFC production immediately at 1986 levels and to reducing total CFCs leading into the twenty-first century. Consumption of the major CFCs was to be frozen at their 1986 levels by mid-1989, reduced to 80 percent of 1986 levels by mid-1993, and reduced to 50 percent of 1986 levels

by 1998. Other halogenated compounds were to be frozen at 1986 consumption levels in 1992.

The protocol was amended in 1990 when new scientific evidence suggested that ozone was being depleted above Antarctica more dramatically than previously assumed. Measurements above Antarctica showed that ozone concentrations declined to between 50 and 95 percent of 1979 levels during certain times of the year. Improved atmospheric models also suggested that the goals of ozone protection could be met only through more stringent curbs on CFC production. The amended Montreal Protocol called for a total phaseout of specified CFCs, halons, and carbon tetrachlorides by 2000 and methyl chloroform by 1995. It also accelerated the rate at which the phaseout would be conducted for CFCs in general.

## Milestones Leading to the Montreal Protocol

| YEAR | EVENT |
| --- | --- |
| 1930's | Chlorofluorocarbons (CFCs) are developed by Du Pont as a safe alternative to toxic refrigerants. |
| 1970 | James Lovelock's electron capture detector reveals the accumulation of CFCs in the atmosphere. |
| 1974 | Mario Molina and Frank Sherwood Rowland show that CFCs degrade through photodecomposition and release ozone-depleting chlorine molecules. |
| 1978 | Pressure from environmentalists causes the United States to ban CFCs as an aerosol propellant; however, worldwide use continues to grow. |
| 1981 | The United Nations Environment Programme forms the Ozone Group and discusses a global treaty to protect the ozone layer. |
| 1985 | Scientists observe a hole in the ozone layer over Antarctica that is linked to stratospheric chlorine. |
| 1986 | Major CFC producers advocate international efforts to limit the growth of CFC emissions. |
| 1987 | The Montreal Protocol is signed; 46 nations, including the United States, commit to a plan to reduce and eventually eliminate CFC production. |
| 1989 | The Montreal Protocol enters into force; signatory nations are required to begin the phaseout of CFCs. |
| 1990 | Revised estimates of ozone depletion lead to a call to cease CFC production by 2000. |
| 1993 | Scientists detect a measurable reduction in atmospheric CFCs. |

Periodic amendments to the protocol since 1990 (in 1992, 1995, 1997, 1999, and 2007) have further accelerated the phaseout schedules. The amendments have also added controlled substances to the protocol, including hydrobromofluorocarbons (HBFCs), hydrochlorofluorocarbons (HCFCs), and bromochloromethane. In 2008, the United Nations Environment Programme estimated that if compliance with the protocol continues, the Arctic ozone layer should return to its pre-1980 level by 2050; the Antarctic ozone layer is expected to recover by 2060-2075.

The Montreal Protocol was a political innovation because it called for a gradual reduction in CFC production and allowed for adjustments in the activities of each treaty member that were flexible enough to respond to new scientific information. A total ban on CFCs would have been unworkable; without reasonably inexpensive alternatives, the distribution of temperature-sensitive medical supplies such as blood and food shipments would have been imperiled because CFCs were essential to most refrigerating units. Many workplaces dependent on air-conditioning also would have been adversely affected. In addition, considerable amounts of industrial machinery with productive lifetimes of twenty to thirty years would have become obsolete immediately.

Considerable disagreement still exists regarding the extent of ozone depletion and its effects. Debate continues over whether the economic cost of finding alternatives to CFCs is outweighed by the estimated increase in skin cancer caused by a global reduction in ozone. Scientists have found evidence that ozone depletion can lead to climate change, reduced productivity in Antarctic waters, and decreased reproduction in amphibians worldwide. Important arguments among the initial signers of the Montreal Protocol concerned the level of production cuts that would be required to end the problem of ozone depletion (one reason the protocol was amended in 1990) and the level of support to which developing nations were entitled for complying with the protocol and forgoing the benefits of cheap CFCs (particularly for refrigeration) that developed countries enjoyed.

*Mark Coyne*

FURTHER READING

Barrett, Scott. "The Montreal Protocol." In *Environment and Statecraft: The Strategy of Environmental Treaty-Making.* New York: Oxford University Press, 2003.

Benedick, Richard. *Ozone Diplomacy: New Directions in Safeguarding the Planet.* Enlarged ed. Cambridge, Mass.: Harvard University Press, 1998.

Kaniaru, Donald, ed. *The Montreal Protocol: Celebrating Twenty Years of Environmental Progress—Ozone Layer and Climate Protection.* London: Cameron May, 2007.

Susskind, Lawrence. *Environmental Diplomacy: Negotiating More Effective Global Agreements.* New York: Oxford University Press, 1994.

Zerefos, Christos, Georgios Contopoulos, and Gregory Skalkeas, eds. *Twenty Years of Ozone Decline: Proceedings of the Symposium for the Twentieth Anniversary of the Montreal Protocol.* Dordrecht, the Netherlands: Springer, 2009.

# Mount Tambora eruption

CATEGORIES: Disasters; atmosphere and air pollution; weather and climate

THE EVENT: Massive volcanic eruption in the East Indies that sent such large amounts of ash and sulfur dioxide into the atmosphere that global temperatures were depressed in the following year

DATES: April 5-12, 1815

SIGNIFICANCE: The Mount Tambora eruption is the most dramatic and unequivocal historical example of the widespread effects of volcanic eruptions on world climate. The main lesson of this event, however, is that even an eruption of such magnitude is unlikely to produce long-lasting climatic perturbation in the absence of additional reinforcing factors. Although the year was long remembered as one of great hardship, no severe famine or political upheaval occurred in the affected areas in 1816-1817.

On April 5, 1815, Mount Tambora, an enormous volcano on the island of Sumbawa, abruptly came to life. Over the next several days the eruption intensified, culminating in a cataclysm on April 12 that pulverized and sent into the atmosphere some 100 cubic kilometers (24 cubic miles) of mountaintop, approximately five times that ejected by Krakatoa in 1883. Pyroclastic flows coursed down the flanks of Mount Tambora, incinerating everything in their path and triggering tsunamis that devastated the coasts of Lombok, Bali, and Sulawesi. Violent eruptions continued through July of 1815. The eruptions and tsunamis killed upward of 40,000 people outright, and an estimated 107,000 died of famine and disease on Sumbawa, Lombok, and Bali owing to the destruction of crops by a thick blanket of volcanic ash. Tambora ranks high among purely natural disasters in terms of immediate loss of life.

The Tambora eruption was the most massive and violent volcanic event in the preceding 10,000 years. (The Mount Mazama eruption, which took place around 5500 B.C.E., was of the same magnitude overall but occurred in three distinct phases over a period of 100 years.) The force of the Tambora eruption propelled fine ash and sulfur dioxide more than 25 kilometers (15.5 miles) into the atmosphere, where it was carried around the globe, blocking incident solar radiation. Observers in Europe and North America noted spectacular sunsets and dirty snow in 1815.

Temperatures in the fall and winter of 1815 remained near normal for the decade 1810-1820, which was unusually cold. Climatologists who have examined snow cores suspect that an as-yet-unidentified volcanic eruption somewhere in the Tropics in 1808 started a cooling trend that intensified the effects of Tambora. May, 1815, was unusually cold in both Europe and North America, delaying the planting of crops, and killing frosts occurred in the northeastern United States in June, July, August, and September, destroying sensitive crops as soon as they were planted. In northern and central Europe, low temperatures and high rainfall depressed the yields of cereal grain crops and effectively prevented hay harvesting. In economies entirely dependent on animal power, the failure of feed crops was a major disaster.

Warm temperate and tropical regions were less affected. Difficulties in distributing the food crops produced in these regions highlighted deficiencies in transportation, especially in North America, where people living inland suffered great privation despite adequate food supplies in the southern coastal areas. The experience of 1816—often called the year without a summer—helped stimulate the building of the Erie Canal and accelerated westward migration as people abandoned farms in New England.

People experiencing the nonsummer of 1816 and the bitter winter of 1816-1817 were unaware of the volcanic connection, although Benjamin Franklin had suggested a possible link earlier. Some ascribed the lack of usual summer weather to very conspicuous sunspots, and one Italian scientist confidently predicted that the sun was about to fail. The bizarre weather sparked a number of notable literary works,

including Lord Byron's "Darkness" (written in July, 1816) and Mary Wollstonecraft Shelley's *Frankenstein* (first published in 1818). Hard times associated with crop failures and the end of the Napoleonic wars produced riots and civil disturbances in Great Britain and elsewhere, but these had no lasting impact. Most of the high mortality rate seen in this period resulted from typhus epidemics that raged in Ireland and Central Europe.

*Martha A. Sherwood*

FURTHER READING

Evans, Robert. "Blast from the Past." *Smithsonian* 33 (July, 2002): 52-57.

Grove, Jean N. *Little Ice Ages, Ancient and Modern*. 2d ed. New York: Routledge, 2004.

Stommel, Henry, and Elizabeth Stommel. *Volcano Weather: The Story of 1816, the Year Without a Summer*. Newport, R.I.: Seven Seas Press, 1983.

# National Oceanic and Atmospheric Administration

CATEGORIES: Organizations and agencies; atmosphere and air pollution

IDENTIFICATION: U.S. federal agency that conducts scientific investigations concerning the conditions of the world's oceans and atmosphere

DATE: Established on October 3, 1970

SIGNIFICANCE: The research and monitoring conducted by the scientists of the National Oceanic and Atmospheric Administration and its many divisions have contributed to a better understanding of the earth's environment.

The National Oceanic and Atmospheric Administration (NOAA), which is part of the U.S. Department of Commerce, is the governmental agency charged with monitoring and conducting research concerning the oceans and the atmosphere. Among the general public perhaps the most widely known division within the NOAA is the National Weather Service, which provides storm warnings, weather reports, and observations of the weather to the public on a daily basis. NOAA's functions, however, include much more than weather-related matters; a variety of divisions within the agency provide advice and scientific information to state and local agencies to aid them in environmental management and policy decisions. In addition to the Weather Service, these are the National Marine Fisheries Service, the National Ocean Service, the Office of Oceanic and Atmospheric Research, the Office of Program Planning and Integration, and the National Environmental Satellite, Data, and Information Service.

The National Ocean Service is dedicated to investigating fresh- and saltwater resources and maintaining their economic and ecological viability. The Office of Oceanic and Atmospheric Research provides support for NOAA's research activities, some of which include investigations into severe storms, El Niño events, and deep-sea thermal vents. The Office of Program Planning and Integration assists NOAA personnel with management and performance issues. The National Environmental Satellite, Data, and Information Service manages data on the earth's ecosystems gathered by the Polar Operational Environmental Satellite and the Geostationary Operational Environmental Satellite, both of which gather a wide range of environmental information.

When NOAA was formed in 1970, it combined three of the oldest agencies in the federal government. The oldest of these, the U.S. Coast and Geodetic Survey, dated back to 1807; established during the presidency of Thomas Jefferson, it was the first agency dedicated to scientific geographic investigation of the coasts and surrounding waters of the United States. The National Weather Service dated back to 1870 and the days of the Army Signal Corps. Ultimately this agency became the National Weather Bureau and then, under NOAA, the National Weather Service. The Commission of Fish and Fisheries, which became the National Marine Fisheries Service, had its beginnings in 1871; it is the nation's oldest agency dedicated to the conservation of food fish.

Under NOAA significant research has been accomplished that has contributed to a better understanding of the earth's environment. For example, NOAA's establishment of systems for monitoring sea temperatures in the South Pacific helped establish reliable El Niño climate predictions. The National Weather Service's programs for tracking severe storms have saved untold numbers of lives. NOAA's coastal monitoring of ocean pollution and its fisheries programs have contributed to environmental policies that support economic growth and sustainability. NOAA's use of

environmental sensing satellite platforms as well as other technologies, such as the Global Positioning System for navigation and the development of modern electronic maritime charts, has also contributed to safer conditions for seafarers.

*M. Marian Mustoe*

FURTHER READING

Artiola, Janick F., Ian L. Pepper, and Mark Brusseau, eds. *Environmental Monitoring and Characterization.* Burlington, Mass.: Elsevier Academic, 2004.

National Research Council. *Environmental Data Management at NOAA: Archiving, Stewardship, and Access.* Washington, D.C.: National Academies Press, 2007.

# Nitrogen oxides

CATEGORIES: Pollutants and toxins; atmosphere and air pollution

DEFINITION: Compounds consisting of nitrogen and oxygen

SIGNIFICANCE: Nitrogen oxides are extremely important environmental pollutants that are closely linked to the energy and agricultural sectors of the global economy. They are causally associated with global climatic change, ozone depletion, acidic precipitation, eutrophication, and photochemical smog formation.

Nitrogen oxides are naturally occurring, biologically active compounds that are produced through a variety of biotic and abiotic processes. Nitrogen-fixing organisms, including mutualists associated with leguminous plants, capture atmospheric nitrogen and produce compounds that are readily converted to nitrites and nitrates by nitrifying bacteria. The biological decomposition of organic matter also leads to the formation of nitrogen oxides. Intense heating causes atmospheric nitrogen and oxygen to react, and lightning is an important naturally occurring, abiotic source of nitrogen oxides in the earth's atmosphere.

The use of fossil fuels and nitrogen-based fertilizers and the cultivation of leguminous plants have substantially increased nitrogen oxide concentrations. Some fossil fuels (such as coal) contain significant amounts of nitrogen, which is oxidized during com-

## U.S. Nitrous Oxide Emissions by Source, 1990-2008

| SOURCE CATEGORY | 1990 | 1995 | 2000 | 2005 | 2006 | 2007 | 2008 |
|---|---|---|---|---|---|---|---|
| Agricultural soil management | 203.5 | 205.9 | 210.1 | 215.8 | 211.2 | 211.0 | 215.9 |
| Mobile combustion | 43.9 | 54.0 | 53.2 | 36.9 | 33.6 | 30.3 | 26.1 |
| Nitric acid production | 18.9 | 21.0 | 20.7 | 17.6 | 17.2 | 20.5 | 19.0 |
| Manure management | 14.4 | 15.5 | 16.7 | 16.6 | 17.3 | 17.3 | 17.1 |
| Stationary combustion | 12.8 | 13.3 | 14.5 | 14.7 | 14.5 | 14.6 | 14.2 |
| Adipic acid production | 15.8 | 17.6 | 5.5 | 5.0 | 4.3 | 3.7 | 2.0 |
| Wastewater treatment | 3.7 | 4.0 | 4.5 | 4.7 | 4.8 | 4.9 | 4.9 |
| Product uses | 4.4 | 4.6 | 4.9 | 4.4 | 4.4 | 4.4 | 4.4 |
| Forestland remaining forestland | 2.7 | 3.7 | 12.1 | 8.4 | 18.0 | 16.7 | 10.1 |
| Composting | 0.4 | 0.8 | 1.4 | 1.7 | 1.8 | 1.8 | 1.8 |
| Settlements remaining settlements | 1.0 | 1.2 | 1.1 | 1.5 | 1.5 | 1.6 | 1.6 |
| Field burning of agricultural residues | 0.4 | 0.4 | 0.5 | 0.5 | 0.5 | 0.5 | 0.5 |
| Incineration of waste | 0.5 | 0.5 | 0.4 | 0.4 | 0.4 | 0.4 | 0.4 |
| Wetlands remaining wetlands | + | + | + | + | + | + | + |
| International bunker fuels | 1.1 | 0.9 | 0.9 | 1.0 | 1.2 | 1.2 | 1.2 |
| Total for United States | 322.3 | 342.5 | 345.5 | 328.3 | 329.5 | 327.7 | 318.2 |

*Source:* U.S. Environmental Protection Agency, *Inventory of U.S. Greenhouse Gas Emissions and Sinks, 1990-2008.* Figures are given in teragrams (1 million metric tons) in carbon dioxide equivalents.

bustion. Fossil-fuel combustion also produces the heat required to oxidize atmospheric nitrogen.

Nitrous oxide is a powerful greenhouse gas with a warming potential three hundred times greater than that of carbon dioxide. Agricultural practices are primarily responsible for increasing nitrous oxide concentrations in the earth's atmosphere, but combustion is also important. The Intergovernmental Panel on Climate Change has identified nitrous oxide as an important driver of human-caused global climate change, and nitrous oxide emissions are internationally regulated under the Kyoto Protocol. Human activities have increased the atmospheric nitrous oxide concentration approximately 18 percent since the dawn of the Industrial Revolution.

Nitrous oxide is also responsible for ozone thinning. Research published in 2009 indicated that nitrous oxide was at that time the most important ozone-depleting compound released by human activities. In the stratosphere, nitrous oxide is converted to nitric oxide, which catalyzes ozone-destroying reactions.

Nitrogen oxides react with moisture in the earth's atmosphere to produce nitric acid, a primary component of acid rain. The most significant environmental damage attributable to nitric acid formation and

acidic precipitation is linked to the combustion of fossil fuels and generally occurs downwind of large cities, power plants, and industrial centers.

Nitrogen oxides are important fertilizers, and their production fosters eutrophication, or the overenrichment of bodies of water with nutrients. Nitrogen-limited estuaries and coastal ecosystems may be particularly sensitive to nitrogen inputs, and atmospheric nitrogen loading is associated with harmful algal blooms and a host of other ecological impacts, including declines in species diversity.

The photodissociation of nitrogen oxides leads to the formation of ozone, and the release of nitrogen oxides from automobiles and stationary sources contributes to the formation of photochemical smog. Exposure to abnormally high ambient nitrogen dioxide ($NO_2$) concentrations may worsen asthma symptoms, and some individuals may run an increased risk of respiratory infection, heart failure, or complications during pregnancy. Average ambient $NO_2$ concentrations have decreased substantially in the United States since 1980.

Fuel-burning appliances, such as gas stoves, furnaces, fireplaces, and space heaters, produce $NO_2$, and indoor $NO_2$ exposure is a recognized public health concern. Research indicates that indoor $NO_2$ can produce respiratory symptoms among children with asthma, even at concentrations well below the U.S. Environmental Protection Agency's ambient air standard of 53 parts per billion.

*Brian G. Wolff*

## FURTHER READING

Jacobson, Mark Z. *Atmospheric Pollution: History, Science, and Regulation.* New York: Cambridge University Press, 2002.

Vallero, Daniel. *Fundamentals of Air Pollution.* 4th ed. Boston: Elsevier, 2008.


# Noise pollution

CATEGORY: Pollutants and toxins

DEFINITION: Harmful or annoying sounds in an environment

SIGNIFICANCE: The problem of noise pollution is particularly acute because noise increases with population density; thus a disproportionately large sector of the human population experiences the adverse effects of exposure to noise. The control of noise requires scientific, social, and sometimes political actions.

Music, speech, and noise are the three basic categories of sound. Noise is simply defined as any unwanted sound. The degree to which a sound is unwanted is, however, a psychological question; the results of exposure to noise may range from moderate annoyance to hearing loss from high volume levels. Furthermore, the interpretation of what constitutes noise is subjective; both music and conversation may be regarded as noise in some places, such as in an office or a library.

## NOISE IN THE ENVIRONMENT

With few exceptions, technological advances from the mid-twentieth century onward have resulted in a steady increase in the amount of unwanted sound. Examples include jet airplanes, automobiles, and ventilation fans. It was thought at one time that human beings should accept and tolerate the noise that went along with the benefits of many industrial advances, but problems associated with exposure to noise have manifested themselves often enough that there is now serious concern about noise pollution in the environment. Noise can be generated in a great variety of ways, but only a few prominent sources of noise emission are part of the daily lives of people in industrialized nations. Noise in the environment can be greatly reduced if these sources of noise pollution can be controlled.

Noise from airplanes poses a major problem in urban areas. Airplanes produce noise through the efflux of jet engines and the high-pitched whine of engine fans. Since 1969 the Federal Aviation Administration (FAA) has legislated acceptable noise levels for commercial airplanes in the United States. Over the years, remarkable engineering innovations have been made in the reduction of jet noise to satisfy the relatively stringent FAA requirements. In the meantime, however, air travel has become increasingly popular. Airlines in the United States have now captured more than 80 percent of all intercity passenger traffic, and the percentage is still on the rise. It is anticipated that the number of takeoffs and landings near major cities will continue to grow throughout the early decades of the twenty-first century. Furthermore, as land prices rise, residential dwellings are encroaching on noise buffer zones near airports in greater numbers. The control of aircraft noise will certainly continue to be pressing in the future.

The deafening din of high-powered trucks and motorcycles is familiar to nearly everyone. Social surveys in cities consistently rank road traffic noise as one of the primary sources of annoyance. Cities can help to control traffic noise by rerouting heavy traffic and smoothing the flow of traffic so that vehicles avoid unnecessary starts, stops, and acceleration. Requirements that motor vehicles be maintained properly can also reduce traffic noise, as can the construction of sound-barrier walls near highways. Because most of the noise from a vehicle traveling at low speed is radiated from the exhaust system, a good muffler is very effective in controlling the acoustical emission. Given the recognition of noise pollution as a serious problem, the U.S. Environmental Protection Agency (EPA) has made several recommendations for reducing noise from motor vehicles.

## INDOOR NOISE

A high proportion of the U.S. workforce is employed in interior environments in which the workers are subject to long periods of exposure to noise. Indoor noise also affects people living in apartments and in houses of relatively light construction. When indoor noise pollution results from sounds produced outside indoor spaces, lining the walls and ceilings of the spaces with acoustic panels and other sound-absorbing materials can be helpful in reducing the noise. In addition, the operation of most home appliances and factory machinery produces noise. Noise from washing machines, drills, and air-conditioning systems arises from friction, unbalanced rotating parts, and air turbulence created by fans. Such noise can be substantially reduced through proper lubrication, the balancing of rotating parts, and the installment of acoustic insulation. In an effort to encourage the design of quieter mechanical products, the American National Standards Institute has published guidelines for manufacturers to use in rating noise emission in their products. Adherence to the guidelines has been less than uniform, but perhaps this will change if consumers and employers express a willingness to pay higher prices for quieter appliances and machinery.

Other sources of noise pollution include some that are not even detectable by the human ear. Ultrasonic and infrasonic noises possess frequencies above and below the audible range, respectively. Such noises are emitted, for instance, through the background hum of high-voltage transmission cables. Although not technically heard, ultrasonic and infrasonic noises affect people in ways similar to audible noise.

ADVERSE EFFECTS

The adverse physiological and psychological effects of noise on people have been the subject of considerable study. Researchers have found that exposure to noise interferes with people's ability to work and to sleep and also infringes on their enjoyment of recreation. Noise pollution has been associated with fatigue, loss of appetite, indigestion, irritation, and headaches.

High-intensity noise has been shown to have adverse cumulative effects on the human hearing mechanism that may produce temporary or permanent deafness. In fact, noise-induced hearing loss has been identified as a major health hazard by the U.S. Department of Labor's Occupational Safety and Health Administration. Noise pollution decreases worker efficiency and increases worker error rates.

*Fai Ma*

FURTHER READING

Berg, Richard E., and David G. Stork. *The Physics of Sound*. 3d ed. Upper Saddle River, N.J.: Prentice Hall, 2005.

Chiras, Daniel D. "Air Pollution and Noise: Living and Working in a Healthy Environment." In *Environmental Science*. 8th ed. Sudbury, Mass.: Jones and Bartlett, 2010.

Kotzen, Benz, and Colin English. *Environmental Noise Barriers: A Guide to Their Acoustic and Visual Design*. 2d ed. New York: Taylor & Francis, 2009.

Nadakavukaren, Anne. "Noise Pollution." In *Our Global Environment: A Health Perspective*. 6th ed. Long Grove, Ill.: Waveland Press, 2006.

Rossing, Thomas D., F. Richard Moore, and Paul A. Wheeler. *The Science of Sound*. 3d ed. San Francisco: Addison-Wesley, 2002.

# North American Free Trade Agreement

CATEGORIES: Treaties, laws, and court cases; atmosphere and air pollution; water and water pollution

THE TREATY: International agreement providing for the removal of trade barriers and reduction of many important legal and financial restrictions among the United States, Canada, and Mexico

DATE: Signed on December 17, 1992

SIGNIFICANCE: The North American Free Trade Agreement is considered by its supporters to mark new directions in international relations by opening the borders of three very different countries in ways that are intended to enhance not only economic but political and cultural relations as well. Although results have fallen short of original expectations, provisions of the North American Agreement on Environmental Cooperation, a side treaty that went into effect at the same time, are intended to raise awareness of environmental protection issues in all three countries.

The North American Free Trade Agreement (NAFTA), an unparalleled trade accord among the United States, Canada, and Mexico, was signed on December 17, 1992. Once ratified by the legislative bodies of the three nations, the agreement went into effect on January 1, 1994. Although the United States and Canada had established open bilateral trading terms in 1988 (the Canada-United States Free Trade Agreement), NAFTA not only aimed at expanding the earlier agreement, particularly in terms of liberalization of conditions for cross-border private investment, but it also sought to integrate terms of trade and investment between two highly developed national economies and a third emerging, or developing, country. An important aspect of Mexico's role in NAFTA was an expectation that the steps toward liberalization of the Mexican economy begun in 1985 (that is, privatization of traditionally state-run companies and increased emphasis on market-oriented economic activity) would continue at a regular pace.

A first and major aim of NAFTA was to eliminate tariff-based trade "barriers" that historically had shielded domestically produced goods, whether agricultural or industrial, from competition from lower-priced foreign imported goods. High tariffs automatically raise prices for imports. An equally important goal was to eliminate, as much as possible, individual national laws hampering the free flow of labor and capital across the signatories' borders.

Beyond measurable economic results, the "spirit" of NAFTA aimed at improving political relations among the three signatories, especially between the United States and Mexico. A very high priority, for example, was (and continues to be) the need for political cooperation in the war against commerce in illicit drugs—increasingly a military as well as a political necessity. Decades-old concerns over the movement of illegal Mexican immigrants across the border into the United States stood to be reexamined in light of NAFTA's commitment to free trade not only in goods but also in cross-border labor arrangements. Another area dependent on mid- to long-term cooperation across national borders involves proposed programs, largely through shared technology, for ecological sustainability.

Numerous research reports appeared before NAFTA went into effect and continued to appear as analysts tried to estimate the relative attractions and disadvantages for one or another of the signatories that might result from the application of the agreement. In fact, during the first decade and a half of NAFTA's operation, certain patterns, some expected and welcomed, others quite controversial, began to take form. Parties that had opposed the agreement seemed convinced that their negative position had been justified. One

argument, for example, was aimed at cross-border investment patterns. Opponents of NAFTA argued that, as U.S. private investment in Mexico increased, levels of investment in the United States itself would be decreased proportionately. Although statistics from the mid-1990's showed some shift in the value of U.S. capital going to Mexico, the overall weight of the movement—not overwhelmingly great—had to be compared with capital going to other, non-NAFTA countries that continue to attract very high levels of U.S. private investment.

On the positive side, supporters of NAFTA could argue that gradual but continuous growth of the Mexican economy would increase its capacity to import a wide variety of goods from its northern neighbors, especially from the United States. The movement of imports into Mexico would obviously be enhanced by reduced Mexican tariff rates—historically as much as three times the tariff rates of the United States. Estimates of rising levels of U.S. exports to Mexico seemed convincing, rising from about twelve billion dollars in 1986 to more than forty billion dollars in 1993.

Perhaps the strongest arguments against trends facilitated by NAFTA have had to do with potentially controversial effects on labor conditions, particularly on both sides of the Mexico-U.S. border. Critics claim that, as more and more companies opt to move their manufacturing activities to locations in Mexico, where labor costs are considerably lower, U.S. factories are losing orders even to the point of having to close down. Another major concern is symbolized by the Spanish term *maquiladora*, which refers to a factory set up in Mexico with the specific aim of importing (duty-free under NAFTA's terms) machinery and parts needed for assembly of a wide range of goods (ranging from clothing to automobiles) to be exported (again duty-free) for sale on the U.S. market. Many *maquiladoras* have been criticized for exploitative labor practices (substandard wages and working conditions, tenuous job security, and so on) that, according to critics, have been ignored by those profiting north of the border.

Support for goals set by NAFTA's founders during the 1990's continued to be voiced by Mexico's political leadership through the first decade of the new century. In an interview aired on American television in March, 2010, for example, President Felipe Calderón stressed his continued belief that a complementarity exists between Mexico's labor-intensive economy and the capital-intensive economies of its northern neighbors, especially the United States. Whether his words reflected a realistic appraisal or hopeful idealism, he concluded that the two nations "need each other."

This shared need is especially clear in matters touching on environmental protection. Although initially there were expectations that NAFTA would create positive conditions for ecological improvements in all three member states, such hopes gradually fell short of realities. On one hand, optimists predicted that hoped-for increases in per-capita income in Mexico as a result of higher levels of capital investment, accompanied by an expected shift away from traditional pollution-intensive industries south of the U.S. border (such as cement and base metals production), would help reduce heavy levels of pollution in Mexico. On the other hand, pessimists claimed that laxity in Mexico's regulation of pollution might even attract heavy polluters from the more developed NAFTA members to the north. The outcomes predicted by either side have not materialized in statistically provable terms.

Tremendously high levels of pollution in and around Mexico City aside (as this problem is beyond the purview of NAFTA technical observers), Mexico's record in dealing with soil erosion, water pollution, and urban solid waste pollution remains controversial. Although NAFTA has from the outset been "armed" with an institution specifically designated to coordinate environmental protection efforts (the Commission for Environmental Cooperation, or CEC, which was created by the North American Agreement on Environmental Cooperation, a side treaty of NAFTA), only one-third of the CEC budget (about nine million dollars in the late 1990's) goes to Mexico. In fact, although Mexico did emerge from the major slump that hit its economy in the 1980's, by 2010 expenditures for environmental improvements (including vital inspections of pollution-intensive industries) had never reached target levels laid down by the CEC.

Many environmental protection frustrations identified in the first decades of NAFTA's existence remained and even grew in the first decade of the twenty-first century. Severe budgetary problems that originally seemed to be solely a Mexican dilemma, for example, affected both Canada and the United States when near collapse of the world financial system struck in 2008-2009. This development raised fears of necessary reductions in funding for environmental projects.

*Byron Cannon*

FURTHER READING

Cameron, Maxwell A., and Brian W. Tomlin. *The Making of NAFTA: How the Deal Was Done.* Ithaca, N.Y.: Cornell University Press, 2000.

Gallagher, Kevin. *Free Trade and the Environment: Mexico, NAFTA, and Beyond.* Stanford, Calif.: Stanford University Press, 2004.

Grinspun, Ricardo, and Yasmine Shamsie, eds. *Whose Canada? Continental Integration, Fortress North America, and the Corporate Agenda.* Montreal: McGill-Queens University Press, 2007.

Hansen, Patricia Isela. "The Interplay Between Trade and the Environment Within the NAFTA Framework." In *Environment, Human Rights, and International Trade,* edited by Francesco Francioni. Portland, Oreg.: Hart, 2001.

Hufbauer, Gary Clyde, and Jeffrey J. Schott. *NAFTA Revisited: Achievements and Challenges.* Washington, D.C.: Institute for International Economics, 2005.

McPhail, Brenda M., ed. *NAFTA Now! The Changing Political Economy of North America.* Lanham, Md.: University Press of America, 1995.

# Odor pollution

CATEGORY: Pollutants and toxins

DEFINITION: Unwanted scents in the environment

SIGNIFICANCE: Human beings are often annoyed by scents they find unpleasant or obnoxious. Although some odors signal the presence of hazardous air pollutants, many are nontoxic; nevertheless, their presence in the environment can have real impacts on quality of life.

Odor complaints are one of the top citizen pollution concerns in many cities in Europe and the United States; about 10 percent of Americans complain about odor pollution problems. The perception of odor pollution involves a combination of cultural expectations and the ability of individuals to perceive various odors. Cultural expectations play a strong role in reactions to particular odors. For example, people who travel outside their home countries may be exposed to olfactory experiences they consider highly unpleasant, while local residents remain oblivious to the same odors.

Malodorous substances and their sources are often difficult to identify and ameliorate. Because the human nose can detect tiny concentrations of certain chemicals, qualitative and quantitative analyses intended to identify malodorous substances are often inconclusive. Thiols rank high on the list of odor pollution complaints, which is not surprising given the widespread industrial applications of these chemicals and the human ability to smell minute concentrations. Ethanethiol is detectable by smell at concentrations as low as 1 part per 2.8 billion parts of air; much higher concentrations are needed for detection by standard chemical tests.

A variety of chemicals emanating from many sources often contribute to odor problems. Establishing a single cause for an odor is difficult because people who are able to identify a particular odor when it is found alone are often unable to identify it as a component in a complex mixture. Another factor is the concentration of a substance, which may influence the acceptability of the odor. This is particularly true of indole, which is classed as pleasant-smelling in minute concentrations but is overwhelmingly unpleasant in moderate and high concentrations. Because indole is a decomposition product of tryptophan, which is used as a chemical reagent and in the manufacture of perfumes and pharmaceuticals, it is a potential odor pollutant in the vicinity of those industrial settings.

Throughout the United States, new housing subdivisions are encroaching into areas that in the past were reserved for agriculture. In such subdivisions, recently transplanted urban dwellers may complain about a range of odors associated with farming activities, such as mowing hay and manuring fields. Solutions to this type of perceived odor pollution may include working to alter the expectations of new residents and recommending that they close their homes and use air conditioners rather than open their windows for "fresh air."

Within cities, odor pollution is often caused by poor methods of disposal of garbage, including food residues. Large urban areas often develop task forces that combine the resources of a pollution-control office, a sewer department, and refuse collectors to seek out and ameliorate the sources of bad garbage odors. Past investigations of vile odors in San Francisco, California, identified the main culprit to be aging butter discharged into the sewer system by restaurants. The odors emanating from the city's sewer system were eliminated by units similar to those used to clean up toxic spills, which removed the source of the stench and sprayed the area with disinfectant.

Many different industries have been identified as contributing to odor pollution, including food processing, paper manufacturing, electric power generation, and waste disposal. Pollution-control officers frequently find that similar manufacturing facilities have different odor problems. Apparently, differences in effluent gases from the facilities may produce differing odor strengths, resulting in different degrees of annoyance among nearby residents.

The kraft paper industry has received considerable attention for its emissions of highly odorous and unpleasant sulfurous gases. In the manufacture of kraft paper, hydrogen sulfide ($H_2S$) is the major odor pollutant. However, the presence of nitric oxide (NO) enhances the unpleasant perception of $H_2S$, which is more readily sensed in an acid gas mixture than in an alkaline gas mixture. Carbonyl sulfide (COS) and sulfur dioxide ($SO_2$) also affect the perceived odor strength of the effluent gases. A kraft operation that controls the acidity of effluent gases may generate fewer complaints about odor pollution than the industry average. Asphalt plants are another notorious odor source. When these plants are fueled by recycled oil, both the energy source and the product may contribute to odor problems. One technique sometimes employed at these facilities is the use of odor-absorbing products to neutralize emissions of hydrocarbons and sulfur dioxide.

An age-old practice for dealing with unwanted or unpleasant odors is to mask them with neutral or pleasant odors; for example, individuals may apply perfumes to cloak perspiration odor, and stores and offices may use "fresh scent" dispensers in their ventilation systems. Some industries have tried releasing masking odors along with known odor pollutants, with dubious success.

During atmospheric inversion conditions, all air pollutants increase in concentration, including odors. Transportation-related odors, including diesel and automobile exhaust, mingle with ozone created during photochemical smog. During these photochemical air-pollution episodes, the acrid odors signal a real public health threat.

*Anita Baker-Blocker*

FURTHER READING
Drobnick, Jim, ed. *The Smell Culture Reader.* New York: Berg, 2006.
Godish, Thad. "Welfare Effects." In *Air Quality.* 4th ed. Boca Raton, Fla.: Lewis, 2004.

Vallero, Daniel. "Effects on Health and Human Welfare." In *Fundamentals of Air Pollution*. 4th ed. Boston: Elsevier, 2008.

# Ozone layer

CATEGORIES: Atmosphere and air pollution; weather and climate

DEFINITION: Region of the lower stratosphere in which most of the earth's ozone is found

SIGNIFICANCE: The earth's ozone layer protects life on the planet from exposure to dangerous levels of ultraviolet light. Because of the introduction of certain human-made chemicals into the atmosphere, the amount of stratospheric ozone steadily declined during the second half of the twentieth century.

Ozone ($O_3$) is a molecule made of three atoms of oxygen. It is considered a trace gas because it accounts for only .000007 percent of the earth's atmosphere. Ozone concentrations are measured in terms of Dobson units, which represent the thickness of all the ozone in a column of the atmosphere if it were compressed—on average only 3 millimeters (0.118 inch) thick. Depending on where it is found in the atmosphere, ozone may have either a positive or a negative impact on life. When ozone is near the earth's surface, it is a major air pollutant, a chief constituent of smog, and a greenhouse gas. Fuel combustion and other human activities increase the quantities of ozone in this atmospheric region. Ozone that resides in the stratosphere protects earth's organisms from lethal intensities of solar ultraviolet (UV) radiation. Without this shield of ozone, which is created through naturally occurring processes, life on earth as it is now known would probably cease to exist. Approximately 90 percent of the earth's ozone is found in the stratosphere.

Ozone concentration peaks in the lower stratosphere, between the altitudes of 20 and 25 kilometers (12 and 16 miles). Within this "ozone layer," two sets of chemical reactions, both powered by UV radiation, continuously occur. In one reaction, ozone is produced; in the other, ozone is broken down into oxygen molecules and ions. Slightly more ozone is produced than is destroyed by these reactions, so that stratospheric ozone is constantly maintained by nature. Because UV radiation is a catalyst for both reac-

tions, most of this radiation is used up and prevented from ever reaching earth's surface.

UV radiation is categorized as UVA, UVB, or UVC, depending on wavelength. UVC rays readily kill living cells with which they come into contact. UVB rays, although less energetic than UVC rays, also damage cells. The ozone shield prevents all UVC rays from reaching the surface; however, some UVB radiation does pass through the ozone layer. It is contact with this radiation that causes sunburns, accelerates the natural aging of the skin, and has been shown to increase rates of skin cancer and cataracts. The American Cancer Society estimates that more than one million new cases of skin cancer occur each year in the United States, mainly as a result of UV radiation. Some researchers estimate that every 1 percent decline in ozone concentration causes a 2 percent increase in UV intensity at the earth's surface, resulting in a greater risk of skin cancer, cataracts, and immune deficiencies.

Increased UV radiation also harms plant and animal life. Some studies have suggested that yields from crops such as corn, wheat, rice, and soybeans drop by 1 percent for each 3 percent decrease in ozone concentration. Increased UV radiation in polar regions impairs and destroys phytoplankton, which makes up the base of the food chain. A decrease in phytoplankton would likely cause population reductions at all levels of the ecosystem. Fewer than 1 in every 100,000 molecules in the atmosphere is ozone, a ratio that both underscores and belies the critical role ozone plays in protecting human health and the global environment.

## DISCOVERY OF THE ANTARCTIC OZONE HOLE

Satellites placed in orbit during the late 1970's allowed scientists to observe concentrations of stratospheric ozone. Observations showed that during the Southern Hemisphere's spring (primarily September and October), the ozone layer above Antarctica thinned dramatically, then recovered during November. These findings were initially dismissed as being caused by instrument error; however, the measurements were confirmed by the British Antarctic Survey in 1985. During that spring, the loss of ozone exceeded 50 percent of normal concentrations. Alarmingly, continued satellite measurements indicated that each year throughout the 1980's and 1990's, the ozone concentration over Antarctica dropped to record lows, and the size of the ozone-depleted area (dubbed the "ozone hole") increased. In 1986 the size of the ozone hole grew larger than the size of the Antarctic continent, and in 1993 the hole was larger than all of North America.

During an intensive Antarctic field program in 1987, extremely high levels of chlorine monoxide (ClO) were found in the stratosphere. This finding was seen by many scientists as evidence that the cause of ozone depletion was chlorofluorocarbons (CFCs), anthropogenic chemicals that make excellent refrigerants, cooling fluids, and cleaning solvents. Since the 1950's, millions of tons of CFCs had been produced in the United States alone. CFC leakage from old refrigerators and automobile air conditioners, combined with a lack of chemical recycling efforts, allowed huge quantities of CFCs to make their way into the atmosphere. Also, one of the CFCs, CFC-11, was used for decades as a propellant in aerosol spray cans until it was banned during the late 1970's.

In the stratosphere, UV radiation breaks down CFCs, causing them to release chlorine (Cl), a gas that readily reacts with ozone. The reaction produces oxygen ($O_2$) and ClO, which then combine to produce $O_2$ and Cl. Thus, at the end of these reactions, the chlorine ion is again free to destroy another ozone molecule. It is estimated that a single chlorine ion may reside in the stratosphere for fifty years or longer and destroy hundreds of thousands of ozone molecules.

Most scientists believed the detection of stratospheric ClO was the "smoking gun" that proved that human-made CFCs were the cause of ozone depletion. However, many industrialists and others opposed to what they perceived as overzealous environmental regulations asserted that the chlorine in the atmosphere was a result of natural processes such as volcanic eruptions or sea spray. This theory was finally debunked in 1995 when scientists also found hydrogen fluoride in the stratosphere. Hydrogen fluoride is not produced by any natural process; however, fluoride would be liberated from a CFC molecule when it is broken down by UV radiation. Scientifically, there is no longer any debate that ozone destruction is a direct result of CFCs. Other chlorine-containing compounds have also been implicated in ozone destruction, as have some bromine-containing compounds. While bromine is less abundant in the stratosphere than chlorine, atom for atom it is more effective than chlorine in destroying ozone.

## GLOBAL OZONE DEPLETION

Since the initial satellite observations of the 1970's and 1980's, the understanding of the complex science of the formation of the Antarctic ozone hole has greatly improved. Most ozone-destroying compounds are released in the Northern Hemisphere; however, they mix throughout the lower atmosphere in about one year and then mix into the stratosphere in two to five years. A key element that enhances ozone depletion is the presence of polar stratospheric clouds. These clouds form only in the polar regions where temperatures in the stratosphere drop to below −80 degrees Celsius (−112 degrees Fahrenheit). The surfaces of these clouds are sites on which inactive chlorine and bromine are converted into the reactive forms that destroy ozone. The dark Antarctic stratosphere becomes so frigid during winter that these clouds are produced in abundance. As spring begins, the stratosphere is hit with solar UV radiation, accelerating the ozone depletion process. A strong vortex of winds circles the South Pole every winter, keeping the Antarctic stratosphere isolated from neighboring air masses. This allows the ozone-destroying reactions to act on a limited amount of ozone that cannot be replenished by ozone from other latitudes. The result is the appearance of the ozone hole each spring. In late spring the winds weaken, allowing comparatively ozone-rich air from other latitudes to mix into the Antarctic atmosphere.

Ozone concentrations declined globally during the second half of the twentieth century as a result of rising levels of atmospheric chlorine and bromine. It was discovered that an ozone hole also forms over the Arctic in the Northern Hemisphere during the spring, although the hole is smaller in size and higher in ozone concentrations than its Antarctic counterpart. The main reason for the difference between the Arctic and Antarctic is that the winds encircling the Arctic are typically not as strong as those in the Antarctic, thus allowing ozone-rich air to mix constantly into the Arctic. Still, significant ozone depletion occurred throughout the 1990's as Arctic spring concentrations dropped about 30 percent below average levels.

Ozone levels in other parts of the world also declined. Ground stations measured decreasing stratospheric ozone levels at midlatitudes. Satellites indicated a negative trend in ozone concentrations in both hemispheres, with a net decrease of about 3 percent per decade. The depletion increased with latitude and was somewhat larger in the Southern Hemisphere. Over the United States, Europe, and Australia, 4 percent per decade was typical. A new threat to ozone followed the eruption of Mount Pinatubo in the Philippines in 1991, the century's second most violent volcanic eruption and the largest one to occur since ozone monitoring began. Massive amounts of sulfur aerosols were injected into the stratosphere. Volcanic aerosols function like the ice crystals in the Antarctic stratosphere by helping convert chlorine and bromine from their inactive state to a reactive one. During the two years following the eruption, record low levels of annual ozone concentrations were recorded: 9 percent below normal between 30 and 60 degrees north latitude.

As expected, measurements also indicated increased UV radiation reaching the earth's surface. During the 1990's, summer levels of UVB increased 7 percent annually over Canada, while winter levels increased 5 percent per year. At the same time, the incidence of skin cancer grew faster than that of any other form of cancer, with reported cases doubling between 1980 and 1998.

## INTERNATIONAL RESPONSE

As evidence grew during the 1980's that CFC production was responsible for ozone depletion, the international community decided to act through a series of treaties and amendments. The multilateral Vienna Convention for the Protection of the Ozone Layer of 1985 outlined the responsibilities of states to prevent human-driven ozone depletion. The Vienna Convention laid the groundwork for the Montreal Protocol on Substances That Deplete the Ozone Layer, which was adopted in 1987. The Montreal Protocol put into place a framework for the phaseout of CFC production in developed countries beginning in 1993. In the following years, as ozone levels kept falling, amendments added more chemicals to the phaseout, accelerated the phaseout timetable, and called for the developing world's participation. By 2010 the Montreal Protocol had been ratified by 196 countries, although not all of them had ratified the subsequent amendments. Under the amended protocol, hydrobromofluorocarbons, CFCs, halons, and carbon tetrachloride have already been phased out; methyl chloroform and methyl bromide have been phased out in developed countries and are scheduled for phaseout in developing countries by 2015; and hydrochlorofluorocarbons are to be phased out by 2020 in developed countries and 2040 in developing countries.

Since the beginning of the twenty-first century, concentrations of ozone-depleting substances in the stratosphere have been on the decline as a result of the Montreal Protocol. These substances have such a long residence time in the atmosphere, however, that even with a ban on production, their concentrations in the stratosphere are not expected to return to pre-1980 levels until 2050 at the earliest. According to the U.S. Environmental Protection Agency (EPA), Antarctic ozone is projected to return to pre-1980 levels between 2060 and 2075. Factors including volcanic eruptions, solar activity, and changes in atmospheric temperature, composition, and air motion could affect the ozone layer's rate of recovery.

Some studies have predicted that maximum ozone depletion will occur between 2000 and 2020, followed by a slow recovery. UV radiation levels at the earth's surface are also expected to increase during the early part of the twenty-first century, bringing with them negative health and ecological effects. The largest ozone hole ever recorded occurred in September, 2006. However, according to an EPA report, over most of the world the ozone layer remained relatively stable between 1998 and 2007.

The story of the ozone layer contains both positive and negative lessons regarding human interaction with the environment. Human production of CFCs and other compounds during the latter half of the twentieth century caused significant, long-term harm to the earth's protective barrier against deadly UV radiation. Human health has been affected, and immeasurable damage has been inflicted on many ecosystems. With the earth in the balance, however, the international community decided to act, and did so forcefully. Without the Montreal Protocol, the fate of the ozone layer and life on earth would have been sealed. It is known that human activity has the power to destroy the global environment, but perhaps human activity may have the power to save it as well.

*Craig S. Gilman*
*Updated by Karen N. Kähler*

## FURTHER READING

Bakker, Sem H., ed. *Ozone Depletion, Chemistry, and Impacts.* Hauppauge, N.Y.: Nova Science, 2009.

Cagin, Seth, and Philip Dray. *Between Earth and Sky: How CFCs Changed Our World and Endangered the Ozone Layer.* New York: Pantheon Books, 1993.

Joesten, Melvin D., John L. Hogg, and Mary E. Castellion. "Chlorofluorocarbons and the Ozone

Layer." In *The World of Chemistry: Essentials*. 4th ed. Belmont, Calif.: Thomson Brooks/Cole, 2007.

Mackenzie, Rob. "Stratospheric Chemistry and Ozone Depletion." In *Atmospheric Science for Environmental Scientists*, edited by C. Nicholas Hewitt and Andrea V. Jackson. New York: Wiley-Blackwell, 2009.

Parson, Edward A. *Protecting the Ozone Layer: Science and Strategy*. New York: Oxford University Press, 2003.

Somerville, Richard C. J. *The Forgiving Air: Understanding Environmental Change*. 2d ed. Boston: American Meteorological Society, 2008.

Turco, Richard P. *Earth Under Siege: From Air Pollution to Global Change*. 2d ed. New York: Oxford University Press, 2002.

United Nations Environment Programme. *Environmental Effects of Ozone Depletion and the Interaction with Climate Change: 2006 Assessment*. Nairobi: Author, 2006.

# Pandemics

CATEGORY: Human health and the environment

DEFINITION: Outbreaks of infectious diseases that affect large numbers of people over large geographic areas

SIGNIFICANCE: Pandemics have long affected human populations around the world, often killing substantial numbers of people, weakening many others, and causing economic distress. Owing to advances in medicine and disease research, some infectious diseases have become less likely to pose pandemic threats, whereas improvements in transportation and increased global travel have increased the likelihood that other diseases may become pandemics.

The spreads of several infectious diseases—most notably plague, cholera, influenza, typhus, smallpox, malaria, and yellow fever—have reached pandemic levels in the past. During the early twenty-first century acquired immunodeficiency syndrome (AIDS), which is caused by the human immunodeficiency virus (HIV), reached pandemic status, with large numbers of cases appearing in all regions of the world and more than 25 million deaths recorded. In 1918-1919 influenza reached pandemic proportions and is estimated to have killed between 20 million and 40 million people worldwide. In 2009 influenza again reached pandemic proportions, although the death rate was much lower than in the earlier pandemic.

## SPREAD OF INFECTIOUS DISEASES

Infectious diseases are either bacterial or viral in nature and spread in different ways. Some, such as smallpox (a virus), are spread through human contact or through material that has come in contact with an infected person. Other diseases, such as influenza (a virus), spread through the air as infected persons cough or sneeze. Bubonic plague (a bacterial disease) is spread to humans through the bites of fleas that have previously bitten infected animals such as rats. (One type of plague—pneumonic plague—is spread through the air from one person to another.) Cholera (a bacterial disease) is spread through drinking water that contains the bacteria *Vibrio cholerae*. The complicated cycle for the spread of malaria (a protozoal disease) involves an infected mosquito biting a person, as is also the case with yellow fever. HIV/AIDS is a viral disease that spreads through contact with bodily fluids of an infected person.

When the global population was small and disaggregated it was difficult for infectious diseases to spread because an infected host might never come in contact with a person susceptible to the disease. The growth of urban areas and improved transportation made it possible to spread infectious diseases readily. If a population has no existing immunity or previous exposure to an infectious disease, a so-called virgin-soil epidemic may result if the population is exposed to the disease. When this occurs, the disease may have a high mortality rate and reach pandemic proportions.

Environmental disturbances such as land clearance, construction of lakes, and climate change may affect the spread of infectious diseases by making it easier for microbes to find hosts. Environmental changes may also expose human populations to new diseases by causing mutations in diseases that enable them to spread from animal hosts to humans, as scientists believe occurred with HIV and some of the hemorrhagic fevers.

As medical knowledge improved during the late nineteenth and early twentieth centuries, it became possible for human beings to prevent the spread of many infectious diseases and to treat and cure those who had been infected. Scientists first identify the agent of a disease, such as mosquitoes for yellow fever and malaria, and then look for ways to control the agent, such as draining standing water to decrease the number of mosquitoes. In some cases it is possible to confer immunity to a disease through vaccination, as occurred with smallpox, which has been eradicated as a disease outside laboratory settings. Other diseases, such as cholera, can be treated with massive re-hydration. Malaria can be prevented with the use of chloroquine. A vaccine exists for bubonic plague, and cases can be treated with large doses of antibiotics.

Since the late twentieth century, drug therapy has been successful in preventing many pandemics. Infectious diseases still occur, but in many cases it is possible for medical personnel to cure such diseases readily and prevent their further spread. However, drug-resistant strains of several diseases, such as malaria and plague, have developed, so that potential for new pandemics exists. HIV/AIDS can be controlled with drug cocktails composed of several drugs, but this viral disease mutates rapidly, so eradication is unlikely. Influenza vaccines exist and are administered every year, but the virus mutates rapidly, so that new vaccines are necessary for every flu season. In some parts of the world, such as Africa, drug therapies are too expensive for most people; this makes diseases such as HIV/AIDS and cholera difficult to combat, enabling them to reach pandemic proportions. Wars and natural disasters can also produce conditions that encourage the spread of diseases such as cholera.

## HISTORICAL OVERVIEW

Smallpox (caused by the virus *Variola major*) is one of the oldest infectious diseases; it is known to have existed in antiquity. During the second century B.C.E. what is known as the Antonine Plague saw smallpox spread throughout the Italian peninsula from the Near East. The third century Plague of Cyprian, which may have also been smallpox, led to massive numbers of deaths in Roman Italy. During the American Revolution smallpox broke out in Boston in 1776 and gradually spread across the continent, nearly decimating several of the Native American tribes of the Great Plains.

Plague (*Yersinia pestis*) is often considered the classic pandemic disease. The first plague pandemic, the Plague of Justinian, in the sixth century, started in Egypt, and from 541 to 750, bubonic plague spread across the Western world, causing Europe's population to drop by as much as 50 percent. From an origin possibly in Mongolia, plague made another appearance in Europe from 1348 to 1350 and then contin-

ued to reappear until the early eighteenth century. The first outbreak, which involved both bubonic and pneumonic plague, killed between 30 and 60 percent of the population in Europe. Combined with other environmental factors, this mortality rate led to a decline in cultivation in some areas and substantially reduced human impacts on the environment.

The third plague pandemic started in western China during the late eighteenth century, spread to coastal cities and India by the late nineteenth century, and reached the United States and South America during the early twentieth century. Although the third plague pandemic did not produce as high a mortality rate as the first two pandemics, it still killed hundreds of thousands of people in China and India. In the early twenty-first century, plague remains endemic in some areas, such as Mongolia, Madagascar, and the American Southwest, raising the possibility of another major outbreak.

Cholera first appeared as a pandemic disease in the early nineteenth century. The disease had apparently existed in India for some time, but it did not reach the level of a pandemic until 1816-1826, when it spread to China, Indonesia, and the Caspian Sea region. Since that time six more cholera pandemics have been recorded. The disease reached England and the United States during the second pandemic (1829-1851). The third pandemic (1852-1860) mainly affected Russia, Japan, and China, and the fourth (1863-1875) affected Europe and Africa. The fifth cholera pandemic (1881-1896) was worldwide, causing at least 250,000 deaths in Europe, 50,000 deaths in the Americas, nearly 300,000 deaths in Russia, and 90,000 deaths in Japan. Advances in public health prevented the sixth pandemic (1899-1923) from having much of an impact on Europe or the United States, but as many as 800,000 people died in India and more than 200,000 in the Philippines. The seventh cholera pandemic, which began in 1962 and continues in the early twenty-first century, introduced a new cholera strain, known as El Tor; this pandemic started in Indonesia and spread to India, parts of the former Soviet Union, Africa, and parts of South America.

It is difficult to identify past influenza pandemics because the symptoms of influenza are similar to those of many other diseases. An influenza pandemic may have affected most of Europe during the late sixteenth century. A major pandemic that may have caused as many as one million deaths spread worldwide in 1889-1890. Some people who contracted the disease and survived had some immunity against influenza during the 1918-1919 pandemic, which appears to have originated in the United States before spreading worldwide, producing particularly high mortality rates in Samoa and Alaska. More than 600,000 people died from the disease in the United States, and it has been estimated that 17 million people may have died from the disease in India.

Influenza reached pandemic proportions again in 1957-1958 and 1968-1969, although it did not produce the large number of deaths that occurred earlier, largely because of improved medical care. Each of these pandemics was caused by a different strain of the influenza virus, which mutates rapidly, making the development of preventive vaccinations difficult. During the early twenty-first century, a new strain of influenza, the H1N1 strain, appeared in China and by 2009 had spread worldwide. Because of improvements in medical care and preventive measures, influenza during the early twenty-first century does not pose the threat of earlier pandemics. Some researchers, however, have voiced concerns that a new "superstrain" of influenza could occur that will be difficult to treat and will cause a high mortality rate.

HIV, the virus that causes AIDS, may have jumped species from monkeys in Africa sometime during the early twentieth century. By the early 1970's AIDS was spreading across the globe, and by 2010 the HIV/AIDS infection rate in parts of sub-Saharan Africa was as high as 25 percent. Some southern and eastern Africa countries have seen declines in population owing to the high mortality rate produced by AIDS as well as related issues. Although HIV/AIDS can be controlled somewhat through drug therapy, this option is very expensive and thus unavailable to most infected people in Africa and Asia. Researchers have estimated that 90 million to 100 million people will die of AIDS in Africa by 2025; projected AIDS deaths in India and China by 2025 are 31 million and 18 million, respectively.

## THE ENVIRONMENT AND PANDEMIC DISEASE

Changes in how human beings relate to the environment have had impacts on the creation and spread of diseases. As humans venture into new environments they encounter new diseases, some of which migrate from other species. In Africa human-made large lakes helped to spread malaria by providing breeding grounds for the mosquitoes that carry the disease, just as controlling standing water all but eliminated the threat of malaria and yellow fever epidemics in the Western world. Climate change, too, may have an impact on the spread of infectious diseases, as warmer temperatures may make it possible for some tropical diseases to spread to new surroundings.

The high death tolls of some pandemics have had impacts on the environment, although it is difficult to trace these impacts clearly. The high death rate among Native Americans produced by contact with European diseases such as smallpox and even measles seems to have depopulated some areas in the Americas in the sixteenth century. The massive death rates produced by the Black Death (possibly a bubonic plague pandemic) in the fourteenth century decreased the impacts of human activities on the environment by a substantial amount. Other pandemics, such as cholera and at times malaria and yellow fever pandemics, have reduced human populations enough to have produced some environmental impacts. The extremely high number of deaths produced by the influenza pandemic of 1918-1919 had little impact on the environment except in isolated regions because of the large size of the world's population by the early twentieth century.

*John M. Theilmann*

FURTHER READING

Barnes, Ethne. *Diseases and Human Evolution.* Albuquerque: University of New Mexico Press, 2005.

Crawford, Dorothy H. *Deadly Companions: How Microbes Shaped Our History.* New York: Oxford University Press, 2007.

Crosby, Alfred W. *America's Forgotten Pandemic: The Influenza of 1918.* 2d ed. New York: Cambridge University Press, 2003.

Engel, Jonathan. *The Epidemic: A Global History of AIDS.* Washington, D.C.: Smithsonian Institution, 2006.

Fenn, Elizabeth A. *Pox Americana: The Great Smallpox Epidemic of 1775-82.* New York: Hill & Wang, 2001.

McMichael, Tony. *Human Frontiers, Environments, and Disease: Past Patterns, Uncertain Futures.* New York: Cambridge University Press, 2001.

Orent, Wendy. *Plague: The Mysterious Past and Terrifying Future of the World's Most Dangerous Disease.* New York: Free Press, 2004.

Theilmann, John M., and Frances Cate. "A Plague of Plagues: The Problem of Plague Diagnosis in Medieval England." *Journal of Interdisciplinary History* 38 (2007): 371-393.

# Pesticides and herbicides

CATEGORY: Pollutants and toxins

DEFINITION: Chemicals designed to kill or inhibit the growth of unwanted organisms

SIGNIFICANCE: Although the use of pesticides, including herbicides, has been beneficial to humankind, enabling increased crop yields and helping to prevent disease, many of the chemicals that have been used to kill pests have had detrimental effects on the environment as well as direct negative effects on human health.

The major types of pesticides in common use are insecticides (to kill insects), nematocides (to kill nematodes), fungicides (to kill fungi), herbicides (to kill weeds), and rodenticides (to kill rodents). Although the use of pesticides has mushroomed since the introduction of monoculture (the agricultural practice of growing only one crop on a large amount of land), the application of chemicals to control pests is by no means new. The use of sulfur as an insecticide dates back before 500 B.C.E. Salts from heavy metals such as arsenic, lead, and mercury were used as insecticides from the fifteenth century until the early part of the twentieth century, and residues of these toxic compounds are still being accumulated in plants that are grown in soil where these materials were used. In the seventeenth and eighteenth centuries, natural plant extracts such as nicotine sulfate from tobacco leaves and rotenone from tropical legumes were used as insecticides. Other natural products, such as pyrethrum from the chrysanthemum flower, garlic oil, lemon oil, and red pepper, have long been used to control insects.

In 1939 the discovery of the utility of dichloro-diphenyl-trichloroethane (DDT) as a strong insecticide opened the door for the development of a wide array of synthetic organic compounds to be used as pesticides. Chlorinated hydrocarbons such as DDT were the first group of synthetic pesticides. Other commonly used chlorinated hydrocarbons include aldrin, endrin, lindane, chlordane, and mirex. Because of the low biodegradability and long persistence in the environment of these compounds, their use was eventually banned or severely restricted in the United States. Organophosphates such as malathion, parathion, and methamidophos replaced the chlorinated hydrocarbons. These compounds biodegrade in a fairly short time but are generally much more toxic to humans and other animals than the compounds they replaced. In addition, they are water-soluble and, therefore, more likely to contaminate water supplies. Carbamates such as carbaryl, maneb, and aldicarb have also been used in place of chlorinated hydrocarbons. These compounds biodegrade rapidly and are less toxic to humans than organophosphates, but they are also less effective in killing insects.

Herbicides, which are used specifically to kill or retard the growth of unwanted plant life, are classified according to the ways in which they work rather than their chemical composition. As their name suggests, contact herbicides such as atrazine and paraquat kill when they come into contact with a plant's leaf surface; these herbicides generally work by disrupting the photosynthetic mechanism. Systemic herbicides such as diuron and fenuron circulate throughout the plant after being absorbed. They generally mimic the plant hormones and cause abnormal growth to the extent that the plant can no longer supply sufficient nutrients to support growth. Soil sterilants such as triflurain, diphenamid, and daiapon kill microorganisms necessary for plant growth and also act as systemic herbicides.

PESTICIDE USE

In the United States, approximately 55,000 different pesticide formulations are available, and Americans apply about 500 million kilograms (1.1 billion pounds) of pesticides each year. Fungicides account for 12 percent of all pesticides used by farmers, insecticides account for 19 percent, and herbicides account for 69 percent. These pesticides are used primarily on four crops: soybeans, wheat, cotton, and corn. By the end of the twentieth century, the annual expenditure on pesticides each year in the United States had reached approximately $5 billion, about 20 percent of which was for nonfarm use. On a per-unit-of-land basis, home owners apply approximately five times as much pesticide to their yards as farmers do to their fields. Worldwide, more than 2.5 tons of pesticides are applied each year. Most of these chemicals are applied in developed countries, but the amount of pesticides being used in developing countries is rapidly increasing. Total annual expenditure on pesticides worldwide exceeds $20 billion, a figure expected to continue increasing as the use of pesticides grows in developing countries.

The use of pesticides has had a beneficial impact

on the lives of humans by increasing food production and reducing food costs. Even with pesticides, insects and other pests reduce the world's potential food supply by as much as 55 percent. Without pesticides, the losses would be much higher, resulting in increased starvation and higher food costs. Pesticides also increase the profit margin for farmers. It has been estimated that for every dollar spent on pesticides, farmers experience an increase in yield worth three to five dollars. Pesticides appear to work better and faster than alternative methods of controlling pests. These chemicals can rapidly control most pests, are cost-effective, can be easily shipped and applied, and have a long shelf life in comparison with alternative methods. In addition, farmers can quickly switch to different pesticides if the pests they are trying to kill develop genetic resistance to a given pesticide.

Perhaps the most compelling argument for the use of pesticides is the fact that pesticides have saved lives. It has been suggested that since the introduction of DDT, the use of pesticides has prevented approxi-

mately seven million premature human deaths from insect-transmitted diseases such as sleeping sickness, bubonic plague, typhus, and malaria. It is likely that even more lives have been saved from starvation because of the increased food production resulting from the use of pesticides. It has been argued that this one benefit far outweighs the potential environmental and health risks of pesticides. In addition, new pesticides are continually being developed, and safer and more effective pest control may be available in the future.

ENVIRONMENTAL CONCERNS

An ideal pesticide would have the following characteristics: It would not kill any organism other than the target pest; it would in no way affect the health of nontarget organisms; it would degrade into nontoxic chemicals in a relatively short time; it would prevent the development of resistance in the organism it is designed to kill; and it would be cost-effective. No pesticide currently available meets all of these criteria, however, and, as a result, a number of environmental

*An American farmworker takes health precautions while preparing pesticides for use on crops.* (USDA)

problems have developed from the use of pesticides. One of these problems is broad-spectrum poisoning. Most, if not all, chemical pesticides are not selective. In other words, they kill a wide range of organisms rather than just the target pest. The extermination of beneficial insects, such as bees, ladybugs, and wasps, may result in a range of problems, including reduced pollination and explosions in populations of unaffected insects.

When DDT was first used as an insecticide, many people believed that it was the perfect solution for controlling many insect pests. Initially, DDT dramatically reduced the number of problem insects; within a few years, however, a number of insect species had developed genetic resistance to the chemical and could no longer be controlled with it. By the 1990's approximately two hundred insect species had genetic resistance to DDT. Other chemicals were designed to replace DDT, but many insects also developed resistance to these newer insecticides. As a result, although many synthetic chemicals have been introduced into the environment, the pest problem is still as great as it ever was.

Depending on the types of chemicals they contain, pesticides remain in the environment for varying lengths of time. Chlorinated hydrocarbons, for example, can persist in the environment for up to fifteen years. From an economic standpoint, this can be beneficial because the pesticide has to be applied less frequently, but from an environmental standpoint, it can be detrimental. In addition, many pesticides degrade in such a way that their breakdown products, which may also persist in the environment for long periods of time, are often toxic to other organisms.

Pesticides may concentrate in animals as they move up the food chain. All organisms are integral components of at least one food pyramid. While a given pesticide may not be toxic to species at the base, it may have detrimental effects on organisms that feed at the apex because the concentration increases at each higher level of the pyramid, a phenomenon known as biomagnification. With DDT, for example, some birds can be sprayed with the chemical without any apparent effect, but if these same birds eat fish that have eaten insects that contain DDT, they lose the ability to metabolize calcium properly. As a result, they lay soft-shelled eggs, which causes the death of most of their offspring.

Pesticides can also be hazardous to human health. Many pesticides, particularly insecticides, are toxic to humans, and thousands of people have been killed by direct exposure to high concentrations of these chemicals. Many of those who have died have been children who were accidentally exposed to toxic pesticides because of careless packaging and storage. Numerous agricultural laborers, particularly in developing countries where there are no stringent guidelines for the handling of pesticides, have also died as a result of direct exposure to these chemicals. Workers in pesticide factories are also a high-risk group, and many of them have been poisoned through job-related contact with the chemicals. Pesticides have also been suspected of causing long-term health problems such as cancer, and some pesticides have been classified as carcinogens by the U.S. Environmental Protection Agency.

*D. R. Gossett*

FURTHER READING

Carson, Rachel. *Silent Spring.* 40th anniversary ed. Boston: Houghton Mifflin, 2002.

Connell, Des W. *Basic Concepts of Environmental Chemistry.* 2d ed. Boca Raton, Fla.: CRC Press, 2005.

Miller, G. Tyler, Jr., and Scott Spoolman. "Food, Soil, and Pest Management." In *Living in the Environment: Principles, Connections, and Solutions.* 16th ed. Belmont, Calif.: Brooks/Cole, 2009.

Monaco, Thomas J., Stephen C. Weller, and Floyd M. Ashton. *Weed Science: Principles and Practices.* 4th ed. New York: John Wiley & Sons, 2002.

Ohkawa, H., H. Miyagawa, and P. W. Lee, eds. *Pesticide Chemistry: Crop Protection, Public Health, Environmental Safety.* New York: Wiley-VCH, 2007.

# Polluter pays principle

CATEGORIES: Treaties, laws, and court cases; atmosphere and air pollution; water and water pollution

DEFINITION: Legal rule that the costs of pollution prevention and remediation should be borne by the entity that profits from the process that causes the pollution

SIGNIFICANCE: The polluter pays principle is designed to impose penalties on parties responsible for producing pollution that damages the natural environment and hence society as a whole. Hence the principle is intended to deter activities that pollute the environment.

The polluter pays principle assigns responsibility for the negative impacts arising from polluting activities to the individuals, companies, and other groups that perform those activities. These negative impacts are designated "social negative externalities." The principle is grounded in both the 1992 Rio Declaration, which came out of the United Nations Conference on Environment and Development (also known as the Earth Summit) and states that national governments should "promote the internationalization of environmental costs and the use of economic instruments," and the International Union for Conservation of Nature's Draft International Covenant of Environment and Development, which states that "parties shall apply the principle that the costs of preventing, controlling, and reducing potential or actual harm to the environment are to be borne by the originator."

The Swedish government may have been the first to apply the polluter pays principle, also known as the principle of extended polluter responsibility, in the 1970's. By moving the responsibility for addressing pollution away from taxpayers and governments to the companies and individuals causing the pollution, the principle makes the costs of waste disposal and pollution remediation part of the cost of the process that produces the pollution. Thus companies and individuals have incentive to reduce the amounts of waste and pollution they produce and to increase their efforts to reuse and recycle materials.

The polluter pays principle is applied differently in different situations. Application of the principle tends to be closely related to market-based incentives to regulate consumption or production activity. These policies are broken down between taxes and tradable permit schemes.

## Radioactive pollution and fallout

CATEGORIES: Nuclear power and radiation; pollutants and toxins

DEFINITIONS: Radioactive pollution is environmental contamination resulting from the release of radioactive materials; fallout is the radioactive particles that fall to earth after detonation of nuclear weapons

SIGNIFICANCE: The use of radioactive materials for civilian and military applications has helped to disseminate radioactivity throughout the environment. Because of the long half-lives of some radioactive pollutants, their detrimental effects on ecosystems and human health can persist for generations and pose major challenges for safe, long-term disposal efforts.

Many people are well aware of the use of radioactive elements in nuclear weapons and reactors. Likewise, the general public understands that nuclear war, nuclear weapons testing, and nuclear reactor accidents such as the disastrous Chernobyl plant meltdown of 1986 can have profound, far-reaching, and long-lasting environmental impacts. It may be less well known, however, that radioactive elements are sometimes used for their physical or chemical properties rather than their radioactivity, and that these less familiar applications can cause small amounts of radioactive pollution. For example, uranium and thorium compounds have been used for centuries to give ceramic glazes brilliant orange and yellow hues. (Although pieces produced in the United States after about 1950 are generally safe to use, it is recommended that other ceramics with uranium or thorium glazes be left for occasional show unless they are tested and known to be safe for use.)

In addition, until the 1980's trace amounts of uranium were used to color porcelain teeth, and tiny amounts of uranium or thorium are present in some tinted contact lenses, eyeglass lenses, and other optical glass. Gas lantern mantles have long used thorium to produce a bright, white glow. Although the hazard to the public is believed to be negligible, nonthorium mantles are now available as an alternative. Tungsten electrodes for arc welding may contain 2 percent thorium for easier starting and greater weld stability. The resulting radiation dose to both the welder and the public is very small. Ionization smoke detectors use tiny amounts of radioactive americium to ionize air in a small chamber. This allows a current to flow. Smoke particles reduce this current and trigger the detector. Americium has a 458-year half-life, so it will remain radioactive long after the smoke detector's service life is over. While a smoke detector poses little threat when installed and used properly, its radioactive content can present a health hazard once the smoke detector is discarded. In order to keep americium out of the waste stream, many U.S. state and local governments conduct roundups of ionization smoke detectors or encourage consumers to return used smoke detectors to their manufacturers.

## INDUSTRIAL, MEDICAL, AND SCIENTIFIC USES

Because small amounts of radioactivity are easily detected, both industries and the medical field use radioactive elements as tracers. For example, radioactive technetium may be injected into a patient's vein so that its progress can be followed with radiation detectors. Imaging systems can then reveal constrictions in the patient's heart or arteries. The technetium used has a six-hour half-life. The nondecayed portion of technetium is eventually eliminated from the body and passes into the sewage system. No special precautions are taken because the radioactivity quickly decays and is greatly diluted with normal waste. Worldwide, more than 30 million diagnostic procedures using various radioactive elements are performed each year.

Radioisotope thermoelectric generators (RTGs) were developed to supplement the power from solar cells or replace them on space missions where sunlight is too weak. The Jupiter mission's *Galileo* spacecraft carried two RTGs, while the Saturn mission's *Cassini* spacecraft carried three. A standard RTG uses the heat from the radioactive decay of 11 kilograms (24 pounds) of plutonium dioxide to produce electricity. Because the process involves no moving mechanical parts, it is very reliable. RTGs are built to withstand the explosion of the spacecraft during launch as well as the heat of reentry.

Although the United States has used RTGs in many space missions, it has had only three accidents involving these power sources. In 1964 a navigational satellite failed to reach orbit and burned up over the Indian Ocean. Its RTG was of an earlier design that also burned up, as was then intended. The resulting plutonium oxide dust settled out of the stratosphere over the next several years. Greatly diluted across the globe, it barely increased background radiation. In 1968 a spacecraft was destroyed after launch by the range safety officer, and its RTGs were recovered intact. In 1970 the lunar module of the damaged Apollo 13 spacecraft reentered the atmosphere over the South Pacific. Its RTG plunged into the Tonga Trench, which is 6 kilometers (3.7 miles) deep. Although it was never recovered, surveys have shown no release of radioactivity.

The Soviet Union used RTGs not only in spacecraft but also as power sources for lighthouses and other navigational beacons in remote locations. Since the breakup of the Soviet Union in late 1991, many of these RTGs have been unattended and have fallen into disrepair. Looters hoping to strip metal parts and sell them to recyclers have stolen unsecured RTGs and received high doses of radiation. Stolen RTGs have been abandoned in forests, dumped in the sea, and in one case left at a bus stop. The United States has taken in interest in helping to decommission Russian RTGs, as they are a source of radioactive material that could be exploited by terrorists.

## DEPLETED URANIUM

Uranium is a widely distributed trace element. Its estimated concentration in the earth's crust is 2.7 parts per million. Pure uranium is a lustrous, silver-white metal, and although it is radioactive, the activity consists of alpha particles and low-energy gamma rays that can be readily shielded to safe levels. The ease with which uranium can be safely handled under favorable conditions has led to its increasing use. Many view this as dangerous.

Natural uranium consists of three different forms, or isotopes: about 0.7 percent is uranium 235, 99.3 percent is uranium 238, and only a trace amount is uranium 234. Weapons-grade uranium must be enriched to at least 90 percent uranium 235, while reactor fuel is generally enriched to 3 to 5 percent uranium 235. Extraction processes leave uranium that has only 0.2 to 0.3 percent uranium 235. This is called depleted uranium. Depleted uranium is only 60 percent as radioactive as natural uranium, but there is a lot of it. The United States entered the twenty-first century with an estimated 480,000 metric tons of depleted uranium; the total world inventory at the turn of the century was more than 1.1 million metric tons.

Rather than provide storage for it as low-level radioactive waste, the U.S. Department of Energy actively seeks uses for depleted uranium. Almost twice as dense as lead, it makes a good radiation shield, and it is used to shield radioactive isotopes shipped to hospitals. Ducrete, a special concrete containing depleted uranium, is used as radiation shielding in shipping casks for spent reactor fuel.

Depleted uranium packs a great deal of weight into a small volume; therefore, it is used to make counterweights for commercial aircraft and the tips of Tomahawk cruise missiles. Its density and hardness led to its use in the armor of the M1A1 Abrams tank and armor-piercing ammunition. Powdered uranium is pyrophoric (that is, it burns spontaneously in air). Some of the depleted uranium in an armor-piercing round burns upon impact with a hard surface. This makes the round more effective against tanks because it helps the round to penetrate armor and often ignites a secondary explosion. While most of the uranium is expected to remain within several meters of the target, a significant amount may not. Some uranium oxide particles formed from burning uranium are smaller than 5 microns and can be carried by the wind for long distances. Such small radioactive particles can also lodge in the lungs of organisms and are potentially hazardous.

Hundreds of metric tons of depleted uranium ammunition have been used in Middle Eastern conflicts such as the 1991 Persian Gulf War and the 2003 invasion and subsequent occupation of Iraq. Depleted uranium ammunition was also employed during the 1992-1995 Bosnia-Herzegovina conflict and the 1999 Balkan war. While depleted uranium rounds are highly effective, their potential impact on the environment and the health of soldiers and civilians makes them a controversial weapon. During the mid-1980's, concerns about possible radiation exposure of ship crews led the U.S. Navy to switch from depleted uranium to tungsten rounds for the Phalanx guns used to defend military vessels against planes and missiles. The long-term health effects of depleted uranium use during wartime have yet to be established. The World Health Organization recommends that monitoring and cleanup operations be carried out in impact zones following military conflict if there is a reasonable possibility that depleted uranium could enter the food chain or groundwater in sufficiently high quantities to pose a health and environmental hazard.

## FALLOUT FROM NUCLEAR WEAPONS

Fallout is the name given to radioactive particles that rain down from the debris cloud of a nuclear explosion. Whether the fallout is local or global depends chiefly on the yield of the weapon. Nuclear weapon yields are measured in terms of how many tons of the high-explosive trinitrotoluene (TNT) would be required to release the same energy. The bomb used in the American attack on the Japanese city of Hiroshima during World War II, for example, had a yield of approximately 13 kilotons. The largest nuclear weapons built by the United States had an estimated yield of 25 megatons (25,000 kilotons).

Yields of less than 100 kilotons produce local fallout, which consists of particles that fall to the ground within twenty-four hours of the explosion. Local fallout is generally most intense near ground zero; however, winds may carry local fallout hundreds of kilometers or more. Larger weapons loft debris higher into the air, and small particles may drift for several days before falling to the ground. Particles lifted into the stratosphere may remain there for months or longer and be carried around the world. The radioactivity of fallout decreases with time, so that the longer it remains aloft, the less dangerous it is. Therefore, local fallout is far more hazardous than global fallout.

When a nuclear weapon explodes, it instantly becomes an expanding fireball of radioactive vapor. The radioactivity chiefly comes from the debris of the nuclear fission of uranium. The explosion also produces a torrent of neutrons that can transform some normal elements into radioactive elements. Wherever the fireball touches the ground, dirt and debris are sucked into the air. As the fireball expands and cools, radioactive vapor condenses into radioactive particles and debris that are pulled into the fireball. Because hot air rises, the fireball rises and forms the hallmark mushroom-shaped cloud. Fallout begins within minutes as the heaviest radioactive pebbles rain down near the stem of the mushroom cloud. Fine particles are carried downwind from ground zero and continue to fall to the earth over the next several hours.

If the wind is steady, the radioactivity of the fallout accumulated on the ground could be described with a series of concentric, elongated ovals. The near ends of all of the ovals would touch at ground zero. The innermost oval would have the highest radioactivity. The next oval outward would mark lower activity, and its far end would extend farther from ground zero. For a 1-megaton bomb, the oval that contains a lethal dose of fallout during the first forty-eight hours would be 1,000 square kilometers (386 square miles) in area. However, fallout radioactivity decays relatively quickly, so that after one year the lethal oval would cover only 1 square kilometer (0.38 square mile). At first, 20,000 square kilometers (7,722 square miles) would be contaminated badly enough that an unprotected person might show signs of radiation sickness within two weeks. After one year, that area would decrease to about 20 square kilometers (7.72 square miles).

Depending on such factors as targeting strategies, timing, and weather, fallout from a large-scale nuclear attack could kill tens of millions of people and endanger hundreds of millions more. Because fallout is radioactive dust, it accumulates most readily on horizontal surfaces such as the ground and the roofs of buildings. If caught in a fallout zone, the best strategy would be to get as far away from the fallout as possible and place as much mass between the person and the fallout as possible. For example, the shelter of a simple basement can reduce the radiation dose to 5 or 10 percent of that of a person in the open.

*Charles W. Rogers*
*Updated by Karen N. Kähler*

FURTHER READING

Bréchignac, François, and Brenda J. Howard, eds. *Radioactive Pollutants: Impact on the Environment.* Les Ulis, France: EDP Sciences, 2001.

Edelstein, Michael R., Maria Tysiachniouk, and Lyudmila V. Smirnova, eds. *Cultures of Contamination: Legacies of Pollution in Russia and the U.S.* Amsterdam: Elsevier JAI, 2007.

Leopold, Ellen. *Under the Radar: Cancer and the Cold War.* Piscataway, N.J.: Rutgers University Press, 2009.

Miller, Alexandra C., ed. *Depleted Uranium: Properties, Uses, and Health Consequences.* Boca Raton, Fla.: CRC Press, 2007.

Till, John E., and Helen A. Grogan. *Radiological Risk Assessment and Environmental Analysis.* New York: Oxford University Press, 2008.

# Radon

CATEGORIES: Atmosphere and air pollution; nuclear power and radiation

DEFINITION: Radioactive gas that occurs naturally in rocks as the decay product of radium

SIGNIFICANCE: Although it accounts for approximately 50 percent of the normal background radioactivity in the environment, radon can pose a health hazard if it accumulates in houses and other buildings.

Unsafe levels of radon have been detected in structures built over soils and rock formations containing uranium. One of the radioactive products of uranium is radium, which decays directly to radon. Every 3 square kilometers (1.2 square miles) of soil to a depth of 15 centimeters (6 inches) contains about 1 gram (0.035 ounces) of radon-emitting radium. Certain regions across the United States and around the world contain comparatively high concentrations of radium in

their rocks and soils. One such area is the Reading Prong, which stretches from southeastern Pennsylvania to northern New Jersey and portions of New York.

Three forms of radon are generated in the decay of uranium in rocks and soils. The potential health risks are posed by the radon isotope with an atomic mass of 222 (radon 222), which has a 3.8-day half-life. Radon 220 and radon 219 also form in rocks and soils, but these isotopes have half-lives of 56 seconds and 4 seconds, respectively. The shorter half-lives of these isotopes compared to radon 222 give them a much greater chance to decay within rocks and soils before they can become airborne; thus they are of lesser radiological significance.

Radon is chemically inert, and within its 3.8-day half-life, the gas can become airborne and enter buildings through small fissures in the foundations. Indoor radon levels are typically four or five times more concentrated than outdoor levels, because air

---

## Radon and Its Link to Lung Cancer

*Breathing low levels of radon, a known carcinogen, can lead to lung cancer. Below are some more facts, provided by the Environmental Protection Agency, about the health risks of radon:*

Lung cancer kills thousands of Americans every year. The untimely deaths of Peter Jennings and Dana Reeve have raised public awareness about lung cancer, especially among people who have never smoked. Smoking, radon, and secondhand smoke are the leading causes of lung cancer. Although lung cancer can be treated, the survival rate is one of the lowest for those with cancer. From the time of diagnosis, between 11 and 15 percent of those afflicted will live beyond five years, depending upon demographic factors. In many cases lung cancer can be prevented; this is especially true for radon.

Smoking is the leading cause of lung cancer. Smoking causes an estimated 160,000 deaths in the U.S. every year (American Cancer Society, 2004). And the rate among women is rising. On January 11, 1964, Dr. Luther L. Terry, then U.S. Surgeon General, issued the first warning on the link between smoking and lung cancer. Lung cancer now surpasses breast cancer as the number one cause of death among women. A smoker who is also exposed to radon has a much higher risk of lung cancer.

Radon is the number one cause of lung cancer among non-smokers, according to EPA estimates. Overall, radon is the second leading cause of lung cancer. Radon is responsible for about 21,000 lung cancer deaths every year. About 2,900 of these deaths occur among people who have never smoked.

dilution occurs in outdoor settings. Contributions to indoor radon levels also come from building materials, well water, and natural gas.

Airborne radon itself poses little hazard to health. As an inert gas, inhaled radon is not retained in significant quantities by the body. The potential health risk arises when radon in the air decays, producing nongaseous radioactive products. These products can attach themselves to dust particles or aerosols. When inhaled, these particles can be trapped in the respiratory system, causing irradiation of sensitive lung tissue. Sustained exposure may result in lung cancer. The U.S. Environmental Protection Agency (EPA) has estimated that more than twenty thousand deaths from lung cancer each year are attributable to radon products. The EPA recommends remediation measures if radon levels in a building exceed 4 picocuries per liter of air. Remediation techniques to relieve indoor radon pollution usually involve ventilating basements and foundation spaces to outside air.

*Anthony J. Nicastro*

FURTHER READING

Hill, Marquita K. "Pollution at Home." In *Understanding Environmental Pollution.* 3d ed. New York: Cambridge University Press, 2010.

McKinney, Michael L., Robert M. Schoch, and Logan Yonavjak. "Air Pollution: Local and Regional." In *Environmental Science: Systems and Solutions.* 4th ed. Sudbury, Mass.: Jones and Bartlett, 2007.

Pipkin, Bernard W., et al. *Geology and the Environment.* 5th ed. Belmont, Calif.: Thomson Brooks/Cole, 2008.

# Rowland, Frank Sherwood

CATEGORY: Weather and climate
IDENTIFICATION: American chemist
BORN: June 28, 1927; Delaware, Ohio
SIGNIFICANCE: Rowland was the first person to discover that chlorofluorocarbons released into the atmosphere were destroying the protective ozone layer, and he was influential in the eventual move to ban production of these compounds.

Frank Sherwood Rowland received his Ph.D. in 1952 from the University of Chicago, where he studied under Willard F. Libby, recipient of the 1960 Nobel Prize in Chemistry for developing carbon-14 dating. Rowland became the first chairperson of the Chemistry Department at the University of California, Irvine, in 1964. During the 1970's, he began to think about expanding his research into new areas. The use of the stable compounds chlorofluorocarbons (CFCs) as refrigerants and propellants in aerosol cans had become widespread during the 1960's. Rowland reasoned that because of their stability, CFCs should persist in the atmosphere for a long time. That led to the question of what happens to them as they accumulate in the atmosphere and enter the protective ozone layer at an altitude of about 12,000 meters (40,000 feet).

About this time, Mario Molina joined Rowland's research group as a postdoctoral fellow and undertook the CFC study. Rowland and Molina quickly discovered that the ultraviolet (UV) light streaming through the atmosphere has enough energy to break a bond in the CFCs, releasing the reactive chlorine atom. Chlorine is known to destroy ozone. Rowland and Molina published a paper in 1974 in the journal *Nature* warning that the increasing use of CFCs could result in the destruction of the ozone layer, allowing damaging UV radiation to reach the earth's surface.

Rowland and Molina had no direct proof that the ozone layer was being destroyed, but they called for a ban on CFC production. The production of CFCs had grown to become a huge industry, making some $8 billion per year, so Rowland and Molina knew that their proposal would be controversial. The Du Pont Corporation was the first and largest producer of CFCs, and it led the defense of the industry. Du Pont argued that such a large industry should not be dismantled based on an unproven hypothesis. As warnings about the possible environmental harm increased, however, public concern led many aerosol packagers to cease using CFCs.

The first evidence of ozone layer destruction came from measurements made by Joseph Farman, an English scientist. Farman's data led to the discovery of the seasonal ozone hole over Antarctica. Satellite data gathered by the National Aeronautic and Space Administration (NASA) confirmed what Farman had reported. The CFC industry continued to resist a ban until 1988, when measurements showed a 6 percent overall decrease in the ozone layer. By this time Du Pont had developed alternatives to CFCs and agreed to cease production. An international agreement reached in Montreal led to a worldwide ban on CFC production effective January 1, 1996. It had been fourteen years since Rowland's initial warnings about ozone destruction. Rowland had not been content to report his scientific results but had actively advocated for the CFC ban before numerous commissions and government committees. For his work, Rowland received the Nobel Prize in Chemistry in 1995 jointly with Molina and Paul Crutzen, who had shown that nitrogen oxides also destroy ozone.

*Francis P. Mac Kay*

FURTHER READING

Hill, Marquita K. "Stratospheric Ozone Depletion." In *Understanding Environmental Pollution.* 3d ed. New York: Cambridge University Press, 2010.

Joesten, Melvin D., John L. Hogg, and Mary E. Castellion. "Chlorofluorocarbons and the Ozone Layer." In *The World of Chemistry: Essentials.* 4th ed. Belmont, Calif.: Thomson Brooks/Cole, 2007.

Miller, G. Tyler, Jr., and Scott Spoolman. "Climate Change and Ozone Depletion." In *Environmental Science: Problems, Concepts, and Solutions.* 13th ed. Belmont, Calif.: Brooks/Cole, 2010.

# Secondhand smoke

CATEGORIES: Human health and the environment; atmosphere and air pollution

DEFINITION: Tobacco smoke emitted from burning cigarettes, cigars, and pipes, as well as that exhaled by smokers

SIGNIFICANCE: A strong respiratory irritant, secondhand smoke has been shown to contain thousands of chemicals, more than two hundred of which are harmful to human health and fifty of which are known cancer-causing agents. When concentrated indoors, secondhand smoke becomes a significant pollutant that greatly reduces air quality.

Since the 1950's research has shown that tobacco smoke poses serious health risks to smokers. The World Health Organization has identified smoking as the single largest preventable cause of disease and premature death. From 1972 onward U.S. surgeons general have cited exposure to secondhand smoke (also called passive or involuntary smoking) as a possible threat to nonsmokers.

In 1986 the U.S. surgeon general reviewed the work of more than sixty physicians and scientists from the United States and elsewhere relating to secondhand smoke exposure. The resulting report, published by the U.S. Department of Health and Human Services and titled *The Health Consequences of Involuntary Smoking: A Report of the Surgeon General* (1986), listed three conclusions: Involuntary smoking causes disease in nonsmokers; children exposed to secondhand smoke have an increased frequency of respiratory infections and respiratory symptoms, as well as a reduced rate of lung function capacity as the lung matures; and separating smokers from nonsmokers sharing the same airspace may reduce the exposure of nonsmokers to secondhand smoke but does not eliminate their exposure.

HEALTH EFFECTS

An examination of the contents of secondhand smoke reveals a complex mixture of more than 4,000 chemicals, at least 250 of which are known to be harmful to human health. These chemicals are present in two forms: gases and particles. The exact chemical makeup of secondhand smoke depends on the type of tobacco burned, the additives it contains, the way it is smoked, and (in the case of cigarettes) the paper in which it is wrapped.

The major gaseous toxins in secondhand smoke include carbon monoxide, carbonyl sulfide, benzene, formaldehyde, hydrogen cyanide, and nitrogen oxides. The presence of carbon monoxide, which is also a component of car exhaust, is a cause for particular concern. When inhaled, carbon monoxide interferes with the blood's ability to carry oxygen to the cells of the body. Red blood cells normally carry oxygen from the lungs to the cells. When carbon monoxide is present, it binds to red blood cells and makes them incapable of taking on the oxygen. The brain, heart, and other tissues do not get the oxygen they need.

The chemicals in secondhand smoke that are in particle form include tar, nicotine, and phenol. Tar is considered to be carcinogenic (cancer causing). Nicotine is toxic, and phenol promotes tumor growth. Many other particulate chemicals in secondhand smoke are carcinogenic, including catechol, benz(a)anthracene, quinoline, arsenic, beryllium, cadmium, chromium, nickel, and polonium 210. So many cancer-causing agents are found in secondhand smoke that the U.S. Environmental Protection Agency, the U.S. National Toxicology Program, and the International Agency for Research on Cancer all classify secondhand smoke

## Health Effects of Secondhand Smoke

*In remarks delivered at a press conference held on the occasion of the publication of the 2006 report* Health Consequences of Involuntary Exposure to Tobacco Smoke: A Report of the Surgeon General, *U.S. surgeon general Vice Admiral Richard H. Carmona emphasized the dangers of secondhand smoke:*

Secondhand smoke is a health hazard for all people: It is harmful to both children and adults, and to both women and men. It is harmful to nonsmokers whether they are exposed in their homes, their vehicles, their workplaces, or in enclosed public places. We have found that certain populations are especially susceptible to the health effects of secondhand smoke, including infants and children, pregnant women, older persons, and persons with preexisting respiratory conditions and heart disease.

It is not surprising that secondhand smoke is so harmful. Nonsmokers who are exposed to secondhand smoke inhale the same toxins and cancer-causing substances as smokers. Secondhand smoke has been found to contain more than 50 carcinogens and at least 250 chemicals that are known to be toxic or carcinogenic. This helps explain why nonsmokers who are exposed to secondhand smoke develop some of the same diseases that smokers do. . . .

We know that secondhand smoke harms people's health, but many people assume that exposure to secondhand smoke in small doses does not do any significant damage to one's health. However, science has proven that there is NO risk-free level of exposure to secondhand smoke. Let me say that again: There is no safe level of exposure to secondhand smoke.

---

as a known cancer-causing agent in humans.

Being in the presence of secondhand smoke produces immediate consequences. Common reactions include burning sensations in the eyes, nose, and throat; increased phlegm; rises in heart rate and blood pressure; and headaches and stomachaches. Exposure to secondhand smoke over time increases the risk not only of lung cancer but also of other lung conditions, heart attacks, and strokes. According to a 2005 report by the California Environmental Protection Agency, secondhand smoke causes approximately 3,400 deaths from lung cancer and 46,000 deaths from heart disease every year among adult nonsmokers. According to a 2006 surgeon general's report, the chances of developing lung cancer increase 20 to 30 percent for a nonsmoker living with a smoker.

Infants and young children are especially susceptible to the dangers of secondhand smoke. Childhood health problems related to smoke exposure range from ear infections to sudden infant death syndrome. Children whose parents smoke in the home experience a significantly higher risk of developing lower respiratory tract infections and asthma. Bronchitis, tracheitis, and laryngitis are three acute respiratory illnesses that children of smokers suffer more frequently than do the children of nonsmokers. Children whose parents smoke also have more difficulty with chronic (persistent and ongoing) coughs and increased phlegm than do the children of nonsmokers. Children who suffer from asthma and are exposed to secondhand smoke show an increase in the severity of their symptoms and the frequency of their attacks.

### EXPOSURE REDUCTION

The social and economic consequences of secondhand smoke are far-reaching. Exposure to secondhand smoke leads to increases in human illness, suffering, and medical expenses. Lost wages caused by illnesses related to secondhand smoke exposure also contribute to the economic costs. The overwhelming body of evidence pointing to the dangers of secondhand smoke for nonsmokers has led to a flurry of public outcry and regulatory action throughout the developed world and, to a lesser extent, among developing nations.

In some countries, measures to reduce nonsmokers' exposure to secondhand smoke limit smoking to designated areas in enclosed public spaces such as airplanes, public transportation vehicles, airports, bars, and restaurants. The obvious problem with such measures—smoke does not confine itself to a designated smoking area—has given rise in other countries to outright bans on smoking in many enclosed public spaces. Concerns regarding the health of such workers as restaurant staff, bartenders, casino staff, and flight attendants, whose jobs require them to spend long hours in smoke-filled environments, has fueled arguments supporting bans.

Many countries have forbidden smoking in all workplaces, including restaurants and bars. In the

United States, smoking is banned on interstate buses, most trains, airplanes flying domestic routes, and most airplanes flying between the United States and international destinations. Facilities in the United States that provide federally funded services to children are required to be smoke-free under the Pro-Children Act of 1994.

Many state and local governments in the United States have banned smoking in public facilities such as schools, hospitals, bus terminals, and airports. Some have also forbidden smoking in restaurants, bars, and other workplaces. Some communities within the United States and in other nations have even declared outdoor public spaces—such as university campuses, stadiums, parks, and beaches—to be official smoke-free zones. Multiple-family dwellings such as apartment buildings and condominiums have become additional targets for smoking bans, as shared ventilation systems make it impossible for residents to smoke without affecting others living in the same complex.

Secondhand smoke exposure can also be reduced through a decrease in the number of smokers. Common methods that governments and private organizations have used to encourage smokers to quit and discourage nonsmokers from starting include conducting antismoking education campaigns, limiting minors' access to tobacco products, increasing the prices of or taxes on those products, and providing health insurance coverage for treatments that help tobacco users quit.

*Louise Magoon*
*Updated by Karen N. Kähler*

FURTHER READING

Boyle, Peter, et al., eds. *Tobacco: Science, Policy, and Public Health.* 2d ed. New York: Oxford University Press, 2010.

California Environmental Protection Agency. *Proposed Identification of Environmental Tobacco Smoke as a Toxic Air Contaminant: Part B, Health Effects.* Sacramento: California EPA Office of Environmental Health Hazard Assessment, 2005.

Dean, Michael. *Empty Cribs: The Impact of Smoking on Child Health.* Frederick, Md.: Arts and Sciences, 2006.

Kluger, Richard. *Ashes to Ashes: America's Hundred-Year Cigarette War, the Public Health, and the Unabashed Triumph of Philip Morris.* New York: Vintage Books, 1997.

U.S. Department of Health and Human Services. *The Health Consequences of Involuntary Exposure to Tobacco Smoke: A Report of the Surgeon General.* Atlanta, Ga.: Author, 2006.

Watson, Ronald R., and Mark Witten, eds. *Environmental Tobacco Smoke.* Boca Raton, Fla.: CRC Press, 2001.

World Health Organization. *Protection from Exposure to Second-Hand Tobacco Smoke: Policy Recommendations.* Geneva: Author, 2007.

# Silver Bay, Minnesota, asbestos releases

CATEGORY: Human health and the environment
THE EVENT: The dumping of mining wastes into Lake Superior by the Reserve Mining Company
DATE: Begun in 1947
SIGNIFICANCE: Mining wastes deposited directly into Lake Superior were found to be the source of asbestos fibers in the drinking-water supply of Duluth, Minnesota, and other nearby communities.

Taconite is a low-grade iron ore used in the making of steel products. Rocks are crushed and the ore is magnetically removed; the residual materials, or tailings, are industrial waste. In 1947 the state of Minnesota, in an attempt to revive mining in the region known as the Iron Range, granted the Reserve Mining Company of Silver Bay, Minnesota, permission to dump the tailings from its taconite processing plant directly into Lake Superior. The company began depositing the tailings into a large chasm in the lake at the rate of 67,000 tons per day.

On June 14, 1973, the U.S. Environmental Protection Agency (EPA) announced that high concentrations of asbestos fibers had been found in the drinking water of Duluth, Minnesota. The fibers were identified by Irving Selikoff, director of an environmental sciences laboratory in New York, as amosite, the same asbestos fibers that, when inhaled, are known to cause lung, stomach, and colon cancers after an incubation period of twenty to thirty years. Duluth's drinking water was coming directly from Lake Superior, which was thought to have the purest lake water in the world.

State and U.S. government experts charged that the Reserve Mining Company's tailings were the source of these fibers and that lake currents had brought them into the water supply of Duluth and sur-rounding communities. At the time of the EPA warn-ing, court action was under way to force Reserve to cease dumping its slurry into Lake Superior. Com-pany executives protested that the fibers had come into the lake naturally from eroding rocks located in tributary streams. They also contended that the "wa-ter scare" was a ploy to influence the court's decision.

With Reserve threatening to close down its mining operation if it were to be forced to find another dump site, Silver Bay residents, nearly all of whom owed their jobs to Reserve, strongly defended the company. Duluth's residents, while wanting Reserve to stop its dumping, generally reacted with equanimity when confronted with the EPA report, many noting that they had been drinking Lake Superior water for more than thirty years without ill effects. As no filtration sys-tem existed that could remove the asbestos fibers from the water, most Duluthians had no option ex-cept to continue to drink the water. Moreover, it was unclear whether ingesting the fibers might have the same effect as inhaling them.

Numerous mortality studies were initiated by Selikoff and other scientists in 1973 to evaluate the impact of the fibers on Duluth's population. These studies failed to confirm that the water posed any seri-ous threat to those who drank it. By 1975 the water scare had dissipated, and soon thereafter the fibers in the water ceased to be an issue. After years of litiga-tion, Reserve agreed to stop dumping tailings in the lake and instead dispose of them on land. The com-pany constructed a basin about 11 kilometers (7 miles) inland from Silver Bay to contain the tailings. In 1990 Reserve announced that it was bankrupt, and the company closed. Subsequently, another mining company reopened the Silver Bay taconite plant and continued to use the tailings basin built by Reserve.

*Ronald K. Huch*

FURTHER READING

Bartrip, Peter. *Beyond the Factory Gates: Asbestos and Health in Twentieth Century America.* New York: Con-tinuum, 2006.

Farber, Daniel A. "Economics Versus Politics." In *Eco-pragmatism: Making Sensible Environmental Decisions in an Uncertain World.* Chicago: University of Chi-cago Press, 1999.

# Smog

CATEGORY: Atmosphere and air pollution

DEFINITION: Air pollution resulting from the combi-nation of smoke with fog or from sunlight acting on unburned hydrocarbons emitted from automo-biles

SIGNIFICANCE: Severe smog episodes have been re-sponsible for many deaths and widespread illness in cities around the world. Growing recognition of the detrimental health effects of smog have led many governments to pass laws designed to reduce chemical pollutants in the air.

Originally a blend of the words "smoke" and "fog," the term "smog" was coined to describe the se-vere air pollution that results when smoke from facto-ries combines with fog during a temperature inver-sion. As one ascends upward from the earth's surface, the air temperature drops by about 3 degrees Celsius (5.5 degrees Fahrenheit) every 300 meters (1,000 feet). Temperature inversions occur when this nor-mal condition is reversed so that a blanket of warm air is sandwiched between two cooler layers. A tempera-ture inversion restricts the normal rise of surface air to the cooler upper layers, in effect placing a lid over a region. When the air above a city cannot rise, the air currents that carry pollutants away from their sources stagnate, causing pollution levels to increase drasti-cally. A combination of severe air pollution, pro-longed temperature inversion, and moisture-laden air may result in what has been termed "killer fog."

KILLER FOGS

Several acute episodes of killer fog occurred dur-ing the twentieth century. One was in the Meuse Val-ley of Belgium. During the first week of December, 1930, a thick fog and stagnant air from a temperature inversion concentrated pollutants spewing forth from a variety of factories in this heavily industrialized river valley. After three days of such abnormal conditions, thousands of residents became ill with nausea, short-ness of breath, and coughing. Approximately sixty people died, primarily elderly people and persons with chronic heart and lung diseases. The detrimen-tal effects on health were later attributed to sulfur ox-ide gases emitted by combusting fossil fuels; the gases were concentrated to lethal levels by the abnormal weather. The presence of coal soot, combined with moisture from the fog, exacerbated the effect.

A second episode occurred in Donora, Pennsylvania, during the last week of October, 1948. Donora is situated in a highly industrialized river valley south of Pittsburgh. A five-day temperature inversion with fog concentrated the gaseous effluents from steel mills with the sulfur oxides released by burning fossil fuels. Severe respiratory tract infections began to occur, especially in the elderly, and 50 percent of the population became ill. Twenty people died, a tenfold increase in the normal death rate.

A third major episode occurred in London, England, in early December, 1952. At that time, many residents burned soft coal in open grates to heat their homes. When a strong temperature inversion and fog enveloped the city for five consecutive days, Londoners began complaining of respiratory ailments. By the time the inversion had lifted, four thousand excess deaths had been recorded. In this case it was not only the elderly who were affected—deaths occurred in all age categories. During the next decade London experienced two additional episodes: one in 1956, which claimed the lives of one thousand people, and one in 1962, which caused seven hundred deaths. The decline in mortality rates resulted from the restriction of the use of soft coal, with its high sulfur content, as a source of fuel. Sulfur oxide compounds are responsible for causing lung problems during such episodes; therefore, the term "killer fog" has come to be replaced by the more accurate "sulfurous smog."

*The city of Los Angeles is well known for its photochemical smog.* (©David Mcshane/Dreamstime.com)

## PHOTOCHEMICAL SMOG

Photochemical smog, first noticed in the Los Angeles basin in the late 1940's, has been an increasingly serious problem in cities around the world. Moisture is not part of the equation in this type of air pollution, and smoke-belching factories dumping tons of sulfur oxide compounds into the atmosphere are not required. Rather, photochemical smog results when unburned hydrocarbon fuel, emitted in automobile exhaust, is acted upon by sunlight. The Los Angeles basin, hemmed in by mountains to the east and ocean to the west, has a high density of automotive traffic and plenty of sunshine. Varying driving conditions mean that gasoline is never completely consumed by automobile engines; instead, it is often changed into other highly reactive substances. Sunlight acts as an energy catalyst that changes these compounds into the variety of powerful oxidizing agents that constitute photochemical smog. This type of smog has a faint bluish-brown tint and typically contains several powerful eye irritants. The chemical reactions also produce aldehydes, a class of organic chemical best typified by an unpleasant odor.

The complicated chemistry of photochemical smog also produces ozone, which is extremely reactive; it damages plants and irritates human lungs. Because ozone production is stimulated by sunlight and high temperatures, it becomes a particularly pernicious problem during the summer, especially during morning rush hours. Under temperature inversion conditions, the ozone created in photochemical smog can increase to dangerous levels. Ozone is highly toxic. It irritates the eyes, causes chest irritation and coughing, exacerbates asthma, and damages the lungs.

Photochemical smog and ozone are now common ingredients in urban air. Although acute episodes of ozone-induced mortality are rare, concerns have grown about the detrimental long-term consequences of the brief but repetitive exposures to ozone consistently inflicted on commuters. It appears as though no curtailment of the problem will be possible in urban areas in the United States without significant changes in transportation systems, strict limits on growth, and radical alterations in lifestyle, including automobile use.

*George R. Plitnik*

## FURTHER READING

Elsom, Derek M. *Smog Alert: Managing Urban Air Quality.* London: Earthscan Publications, 1996.

Grant, Wyn. *Autos, Smog, and Pollution Control: The Politics of Air Quality Management in California.* Brookfield, Vt.: Edward Elgar, 1995.

Hinrichs, Roger A., and Merlin Kleinbach. *Energy: Its Use and the Environment.* 4th ed. Belmont, Calif.: Thomson Brooks/Cole, 2006.

Jacobs, Chip, and William J. Kelly. *Smogtown: The Lung-Burning History of Pollution in Los Angeles.* Woodstock, N.Y.: Overlook Press, 2008.

Vallero, Daniel. *Fundamentals of Air Pollution.* 4th ed. Boston: Elsevier, 2008.

# Solar radiation management

CATEGORY: Weather and climate

DEFINITION: Subfield of geoengineering concerned with the alteration of the earth's climate through changes in the interaction between sunlight and the planet

SIGNIFICANCE: A small but growing number of scientists have suggested that strategies to limit global warming through reduction of greenhouse gas emissions may not work, either because of lack of willingness to limit emissions or because global warming may have already progressed to the point that it cannot easily be reversed. Solar radiation management has been proposed as a possible countermeasure in the event that limiting greenhouse gas emissions fails to halt global warming.

Most strategies for combating global warming involve reducing the amount of greenhouse gases in the earth's atmosphere. The strategy of solar radiation management, in contrast, involves either limiting the amount of sunlight reaching the earth's surface (solar irradiation mitigation) or reflecting some of the sunlight that does reach the surface back into space (albedo modification). Both approaches would result in less absorption of sunlight and thus less solar heating of the planet. Theoretically, less solar heating could counter the increased atmospheric heating linked with greenhouse gases.

## ALBEDO MODIFICATION

One way of limiting how much sunlight is absorbed to heat the earth is simply to reflect the light back into space. The fraction of sunlight striking a planet that is reflected is called the albedo of the planet. The higher the albedo, the more light is reflected. Variations in albedo occur naturally on local scales. On cloudy days sunlight is reflected into space by clouds, causing the temperature to be lower on the ground. One suggested approach to solar radiation management is to increase the earth's albedo artificially to lower the planet's temperature by reflecting sunlight into space so that it cannot be absorbed.

Several strategies for making modifications to earth's albedo have been suggested. One idea is to mimic the effect of cloudy days by making more clouds or making existing clouds thicker. Cloud seeding can be accomplished through the injection of particulate matter into the atmosphere to form nucleation sites for water droplets. Other suggestions include using ships with powerful pumps to inject a seawater mist into clouds, making them thicker and more reflective.

Very large volcanic eruptions can send sulfur dioxide and other sulfur contaminants high into the stratosphere. These sulfur contaminants can be reflective enough to increase the earth's albedo by several percent. Since the sunlight is reflected in the stratosphere, it does not reach the ground or the lower atmosphere, and the surface temperature of the earth drops. Historically, massive volcanic eruptions have often been followed by years of cooler weather. One idea for increasing the earth's albedo involves injecting large amounts of similar sulfur contaminants into the stratosphere to cause conditions that would mimic those that lead to the temperature reductions that sometimes follow massive volcanic eruptions.

Albedo modifications can be made on land, too. Some scientists have suggested that simply painting roads white and making the roofs of buildings reflective would have a local effect on weather, particularly near large urban areas. It is unclear whether making human construction artifacts more reflective would have a sufficient impact on global climate to compensate for increased greenhouse gases if that technique is not used in conjunction with other measures to limit global warming. More radical proposals have included the suggestion that reflective paint could be applied to deserts, to mimic the increase in albedo of snowfall. Painting the ground white near where ice has melted is one possible strategy to limit the positive feedback of melting polar ice caps.

## SOLAR IRRADIATION MITIGATION

Some volcanic eruptions inject more light-absorbing ash than reflective particles into the stratosphere. This ash, by absorbing light, reduces the sunlight that reaches the earth's surface, in much the same way as reflecting light would do, causing a cooling of the planet's surface. Some scientists have suggested that the injection of soot or even dirty airplane exhaust into the stratosphere could act to cool the earth's surface. This strategy has many drawbacks, however. Particles absorbing solar energy in the stratosphere, although reducing the absorption of solar energy at the ground, still result in solar energy being absorbed somewhere in the atmosphere, possibly resulting in an eventual change in atmospheric structure. This could eventually lead to temporary reduction in the temperature on the ground at the expense of much larger climate disruption years later.

One very ambitious proposal calls for deploying a cloud of mirrors in space between the earth and the sun. Through the deployment of enough objects to reflect less than 2 percent of the sunlight reaching the earth, a thermal balance could be reached, offsetting the effect of increased atmospheric greenhouse gases. Once launched, such a space-based system would require far less effort to maintain than other solar radiation management plans. Reflectors in space are passive systems, so they continue to operate even if they are not actively maintained. The only maintenance required would be to make sure that they stay in stable orbits. This would be achievable if they could be placed in high orbit around the earth, but there they would interfere with spacecraft and the activities of communication satellites. Another possible location for a space-based solar shield would be at the point between the earth and the sun known as the L1 Lagrange point. A gravitational balance between the earth and sun causes objects at the L1 point to tend to stay near that point.

A space-based shield would not raise the concerns associated with the injection of pollutants into the atmosphere to alter the earth's albedo. Furthermore, because such a shield would make no modifications to the atmosphere or to land on the earth, it would not have the possible negative environmental impacts that other approaches to solar radiation management would have. A major difficulty with implementing any plan to create a space-based shield, however, is that it is extremely expensive to launch objects into space.

## CRITICISMS

Numerous criticisms have been leveled against the suggestions that have been made by scientists examining the possibility of solar radiation management. Critics point out that any approach to solar radiation management on a global scale would be exceedingly expensive, would require a great deal of international cooperation, and would require significant investments of time and effort. Even proponents of geoengineering admit that implementation of any of these plans on a global scale would be an extremely difficult and expensive undertaking; they assert, however, that solar radiation management may be the only way to check otherwise unstoppable climate change. There is no agreement among climate scientists on how much climate change is unstoppable without such drastic measures.

Solar radiation management is often seen as only a temporary measure—it may have an effect opposite to that of greenhouse gases, but it does not address the root problem of excessive concentrations of atmospheric greenhouse gases. If solar radiation management programs were to be deployed successfully and then ever stopped for some reason—for example, because of war or the economic collapse of the nations supporting the programs—then the high levels of greenhouse gases in the atmosphere could potentially result in a catastrophic rise in global temperatures. Even proponents of solar radiation management thus often suggest that it be only one part of a multipronged approach to controlling climate change.

A further criticism of solar radiation management is rooted in the fact that solar heating drives many aspects of the earth's ecology. Interfering with natural solar radiation could potentially result in such unintended consequences as changes in the character and frequency of storms, which could result in drought and flooding in diverse areas of the earth. Altering the climate in such a way may also change the acidity of the oceans, resulting in wide-scale extinctions.

Environmentalists often oppose atmospheric modifications such as injecting materials into the stratosphere or into clouds because, they argue, these actions amount to intentional pollution of the planet; the materials that scientists have proposed injecting into the atmosphere could have deleterious effects on the health of humans and other species. Proponents of these plans argue that despite the risks of possible negative health effects, these effects would ultimately be less damaging than the effects of unchecked global warming. Space-based solar radiation management plans would have the fewest potential negative impacts on the earth's environment, but they would also be among the most expensive to implement.

*Raymond D. Benge, Jr.*

FURTHER READING

Angel, Roger. "Feasibility of Cooling the Earth with a Cloud of Small Spacecraft Near the Inner Lagrange Point (L1)." *Proceedings of the National Academy of Sciences* 103, no. 46 (2006): 17184-17189.

Bala, G. "Problems with Geoengineering Schemes to Combat Climate Change." *Current Science* 96, no. 1 (2009): 41-48.

Hoyt, Douglas V., and Kenneth H. Schatten. *The Role of the Sun in Climate Change.* New York: Oxford University Press, 1997.

Kerr, Richard A. "Pollute the Planet for Climate's Sake?" *Science* 314, no. 5798 (2006): 401-402.

Launder, Brian, and J. Michael T. Thompson, eds. *Geo-engineering Climate Change: Environmental Necessity or Pandora's Box?* New York: Cambridge University Press, 2010.

Levi, Barbara Goss. "Will Desperate Climates Call for Desperate Geoengineering Measures?" *Physics Today*, August, 2008, 26-28.

Morton, Oliver. "Is This What It Takes to Save the World?" *Nature* 447, no. 7141 (2007): 132-136.

# Space debris

CATEGORIES: Atmosphere and air pollution; pollutants and toxins

DEFINITION: Human-made nonfunctional objects that are in orbit around the earth

SIGNIFICANCE: Space debris—which consists of nonfunctioning spacecraft, rocket bodies, refuse from missions, and fragments thereof—poses a hazard for space missions, satellite-based services, and people both in space and on earth. As space-faring nations have become more aware of the dangers of this debris, they have worked to minimize its generation during operations in space.

Since 1957 human beings have launched thousands of satellites and other spacecraft. Most of the spacecraft launched successfully achieve orbit. Those that explode after attaining orbit altitude and those that fail after achieving orbit become space debris (also known as orbital debris or space junk). Anything that reaches orbit altitude—about 300 kilometers (186 miles) above the earth's surface—becomes a satellite of the earth. Once in orbit, objects are constantly under the pull of the earth's gravity, and, in time, they slowly fall from orbit. The greater the distance from the earth, the longer an object will remain in orbit. Above 1,000 kilometers (621 miles) objects can remain in orbit for at least a century, objects orbiting at an altitude of 800 kilometers (497 miles) are likely to fall to earth within decades, and those at altitudes between 200 and 600 kilometers (124 and 373 miles) tend to remain in orbit for several years at best.

The U.S. Air Force Space Surveillance Network, which routinely tracks artificial objects orbiting the earth, has cataloged roughly 19,000 debris objects larger than 10 centimeters (4 inches) in diameter. An estimated 500,000 orbiting particles are between 1 and 10 centimeters (0.4 and 4 inches) in diameter. Particles measuring less than 1 centimeter in diameter probably number in the tens of millions. Most of the debris orbits within 2,000 kilometers (1,243 miles) of the earth's surface, with the greatest concentrations accumulating at altitudes between 800 and 850 kilometers (497 and 528 miles).

Each object in orbit runs the risk of running into another object. The volume of space surrounding the earth is immense, and the chances of a collision between two objects are relatively low; however, the likelihood of collision increases when the objects occupy the same orbit. Because certain orbits are particularly desirable for satellites used for communications and surveillance purposes, various nations and commercial interests place their satellites into these positions, thereby increasing the chances of collision.

Many different kinds of space debris orbit the earth. From the 1960's through the mid-1980's, nations deliberately destroyed orbiting satellites while testing weapons for antisatellite warfare. Other forms of space debris have less dramatic origins, such as astronaut Ed White's glove, which slowly drifted away from his Gemini spacecraft in 1965. Each item adds to the ever-increasing number of human-made objects orbiting the earth. Collisions between objects, and explosions of residual fuels in abandoned rocket engines, break existing debris into many smaller pieces.

Objects ranging in size from spent rocket boosters and nonfunctional satellites to small chips of paint, solid-fuel fragments, and coolant droplets have the potential to damage spacecraft. It is not merely the mass of an object that poses a danger but also its high velocity. At orbits below 2,000 kilometers, debris travels at speeds of 7 to 8 kilometers (4.3 to 5 miles) per second, so that even tiny particles can pit space shuttle cockpit windows and damage unshielded satellite components.

## HAZARDS

The space debris population has grown great enough that it has become standard practice to shift unmanned satellites out of harm's way when large debris (objects larger than 10 centimeters) is detected. Space shuttle flights have to adjust course to avoid debris reported by the Space Surveillance Network. The International Space Station is heavily shielded against objects smaller than 1 centimeter, but it has the capability to maneuver away from larger tracked objects.

Only one collision between large, intact satellites has ever occurred. In February, 2009, an operational U.S. Iridium communications satellite accidentally struck a deactivated Russian Cosmos communica-

tions satellite. Both spacecraft were destroyed, and more than 1,500 large fragments were generated. The amount of large debris had already been dramatically increased two years earlier, when in January, 2007, China conducted an antiweapons test in which it used its aging Fengyun-1C weather satellite as a target. The resulting destruction created roughly 2,600 large debris fragments and hundreds of thousands of smaller particles.

Efforts to minimize the problems associated with

## Naturally Occurring Space Material

The term "space debris" is sometimes used to refer to naturally occurring material as well as that generated by human activity. However, as space technology consultant Mark Williamson notes in *Space: The Fragile Frontier* (2006), natural space objects such as meteoroids, meteorites, and cosmic dust are inherent to the space environment, unlike anthropogenic material. Humankind can guard against the flux of natural space material but cannot halt or lessen it; humans can increase it, however, by reducing a single larger body into countless smaller fragments.

An estimated 25 million bits of natural space material collide with the earth each day, the majority of it in the form of cosmic dust. These particles are so small that they do not even appear as meteors as they pass into the atmosphere. Most meteors that are seen are particles the size of a pea, and sometimes the larger ones reach the earth's surface as meteorites.

Occasionally an asteroid-sized object or comet collides with the earth. Such collisions have global implications. The impact destroys the comet and forms a huge crater. An enormous amount of gas and dust is carried into the atmosphere, creating a blanket of debris that blocks sunlight. This begins a "nuclear winter"-type effect, which can last anywhere from a few months to several years. During this time most life-forms will die as a result of the disruption of the food chain. Many scientists believe that such an event led to the mass extinctions of dinosaurs about 65 million years ago.

Although giant impacts are one cause of "nuclear winter," cosmic dust can produce the same effect. The solar system periodically runs into a cosmic dust cloud, thereby dramatically increasing the amount of dust that enters the atmosphere. A similar situation also results from periodic meteor storms.

As the earth runs into the debris of old comets, the number of meteors that enter the atmosphere increases. The earth occasionally encounters a particularly dense region of comet debris. During such meteor storms, thousands of meteors can be seen each hour. The most notable is the Leonid meteor storm, which occurs every thirty-three years. Many scientists fear that increases in comet debris could knock out hundreds of satellites as they are hit by microscopic particles and greatly affect global positioning and communication capabilities.

space debris include the boosting of geostationary satellites that have ended their missions out of their orbits (near 36,000 kilometers, or 22,369 miles, above the earth's surface) into a higher "disposal orbit." Similarly, deactivated satellites that operated at lower altitudes may be moved to even lower orbits that will decay more quickly, hastening the satellites' fall to earth. If a satellite fails to burn up in the atmosphere, however, it can present a threat to people and property on the earth's surface.

In 1978 a Soviet satellite with a nuclear power source survived reentry and strewed small amounts of radioactive material across Canada. The following year, large pieces of the Skylab space station withstood a fiery plunge through the atmosphere and scattered debris across western Australia. In 2001 a rocket upper stage that had been part of a 1993 global positioning satellite launch fell to earth in the Saudi Arabian desert. All of these incidents would have caused considerable damage if the debris had not landed in sparsely populated areas. Only one instance has been recorded of a person being struck by space debris: In 1997 a bit of woven metallic material from a Delta II rocket fuel tank hit an Oklahoma woman on the shoulder but did not injure her. On average, one piece of cataloged space debris falls out of orbit every day, usually burning up in the atmosphere.

### MITIGATION MEASURES

Careful design and operational measures can keep new space missions from contributing unnecessarily to the proliferation of space debris. For example, upper stages of launch vehicles can be placed at lower altitudes so that their orbits decay sooner. Since 1988 the United States has had an official policy of minimizing debris from governmental and nongovernmental operations in space, and the U.S. government approved a set of standard practices for the mitigation of space debris in 2001. The governments of France, the European Union, Japan, and Russia also have issued guidelines pertaining to space debris. Additional guidelines have been published by the United Nations Committee on the Peaceful Uses of Outer Space (COPUOS) and the Inter-Agency Space Debris Coordination Committee (IADC), a group established by the world's leading space agencies in 1993.

Cleaning up existing space debris remains an expensive and technologically challenging prospect. Proposed solutions have included hastening objects' fall to earth by using lasers to slow their orbits and conducting special robotic space missions to grab and haul debris. Solutions that are both technically feasible and economically viable have yet to be developed. In addition, the development of technologies for the cleanup of space debris is controversial because any methods capable of moving spacecraft have potential weapons applications.

A 2006 study sponsored by the U.S. National Aeronautics and Space Administration (NASA) Orbital Debris Program concluded that, if no new launches were conducted and no new objects introduced to earth's orbit, the number of objects falling out of orbit over the next half century would balance the number of new objects created through collisions. After 2055, however, the increasing number of collision-generated fragments—which would go on to create their own catastrophic collisions—would overtake the number lost through decaying orbits.

The international space community's concern in the wake of the 2006 NASA study, China's 2007 weapons test, and the 2009 satellite collision led to the first International Conference on Orbital Debris Removal, convened in December, 2009. Participants examined the many technical, economic, legal, and policy issues surrounding near-earth space cleanup, but they reached no conclusions regarding exactly how humankind might best address the worsening problem of space debris.

*Paul P. Sipiera*
*Updated by Karen N. Kähler*

### FURTHER READING

Inter-Agency Space Debris Coordination Committee. *IADC Space Debris Mitigation Guidelines.* Vienna: United Nations, 2002.

Johnson, Nicholas L., and Darren S. McKnight. *Artificial Space Debris.* Updated ed. Malabar, Fla.: Krieger, 1991.

Klinkrad, Heiner. *Space Debris: Models and Risk Analysis.* New York: Springer, 2006.

National Research Council. *Orbital Debris: A Technical Assessment.* Washington, D.C.: National Academy Press, 1995.

Simpson, John A., ed. *Preservation of Near-Earth Space for Future Generations.* 1994. Reprint. New York: Cambridge University Press, 2006.

Smirnov, Nickolay N., ed. *Space Debris: Hazard Evaluation and Mitigation.* London: Taylor & Francis, 2002.

United Nations Committee on the Peaceful Uses of

Outer Space. *Technical Report on Space Debris.* New York: United Nations, 1999.

Williamson, Mark. *Space: The Fragile Frontier.* Reston, Va.: American Institute of Aeronautics and Astronautics, 2006.

# Stockholm Convention on Persistent Organic Pollutants

CATEGORIES: Treaties, laws, and court cases; pollutants and toxins; human health and the environment

THE CONVENTION: International agreement banning or severely limiting the manufacture and use of certain substances linked to neurological, reproductive, and immune system damage in people and animals

DATE: Opened for signature on May 23, 2001

SIGNIFICANCE: The Stockholm Convention on Persistent Organic Pollutants represents an international effort to reduce the threat of many persistent organic pollutants, which have been linked to cancer, birth defects, and other neurological, reproductive, and immune system damage in people and animals. At high levels, these chemicals can damage the central nervous system, and many also act as endocrine disrupters, causing deformities in sex organs as well as long-term dysfunction of reproductive systems.

The United Nations Environment Programme's Governing Council in 1995 identified twelve persistent organic pollutants (POPs) as the subjects of an eventual ban on manufacture and use worldwide because these substances, known as the dirty dozen, damage the ecosphere and the diversity of life supported by it. While many of these pollutants had already been banned in the United States, other countries had continued to manufacture and use them. The international ban was negotiated in Stockholm late in 2000, and ratification of the protocol followed in 2001.

The substances that were targeted for elimination by the Stockholm Convention included the organochlorine pesticides (such as chlordane, mirex, hexachlorobenzene, endrin, aldrin, toxaphene, heptachlor, and dichloro-diphenyl-trichloroethane, or DDT) and industrial chemicals (including polychlorinated biphenyls, or PCBs, and the supertoxic dioxins and furans). DDT was allowed limited use because no other inexpensive alternatives were available to combat the mosquitoes that spread malaria. Because some of these substances (notably the PCBs and dioxins) are actually families comprising hundreds of chemicals, they could just as aptly be called the dirty hundreds as the dirty dozen.

Synthetic organochlorines such as dioxins and PCBs are perfect vehicles for worldwide pollution because they ignore boundaries, natural or artificial. These chemicals also bioaccumulate (biomagnify, or intensify in potency) along the food chain, sometimes to thousands of times their original toxicity, posing special perils to animals, including human beings, who eat meat and fish. Problems related to their toxicity are especially acute in places, such as the polar regions, where currents in the atmosphere and oceans cause organochlorines to accumulate. Organochlorines produced in the past for commerce and those created unintentionally can be found in the air and in lakes, oceans, soils, sediments, and animals, including humans, in every region of the planet.

POPs are not soluble in water, but they dissolve easily in fats and oils, accumulating in the bodies of living organisms and becoming more concentrated as they move along the food chain. Extremely small levels of such contaminants in water or soil can magnify into lethal hazards to predators who feed at the top of the food web, such as dolphins, polar bears, herring gulls, and human beings.

In some regions of the world, indigenous peoples whose diets consist largely of sea animals (whales, polar bears, fish, and seals) have long been consuming a concentrated toxic chemical cocktail. Abnormally high levels of dioxins and other industrial chemicals have been detected in Inuit mothers' breast milk. People in some villages in the Arctic experience higher concentrations of PCBs than anyone else on earth except victims of industrial accidents. They are at the top of a food chain composed mainly of PCB-laced polar bears, seals, and other animals.

*Bruce E. Johansen*

FURTHER READING

Colborn, Theo, Dianne Dumanoski, and John Peterson Myers. *Our Stolen Future: Are We Threatening Our Fertility, Intelligence, and Survival? A Scientific Detective Story.* New York: Penguin Group, 1996.

Cone, Marla. *Silent Snow: The Slow Poisoning of the Arctic.* New York: Grove Press, 2005.

Johansen, Bruce E. *The Dirty Dozen: Toxic Chemicals and the Earth's Future.* Westport, Conn.: Praeger, 2003.

# Sudbury, Ontario, emissions

CATEGORY: Atmosphere and air pollution

IDENTIFICATION: Release of sulfur dioxide into the air by industry in Sudbury, Ontario, Canada

SIGNIFICANCE: After sulfur dioxide released during the processing of nickel and copper ore caused vast ecological devastation in the area in and around Sudbury, an environmental cleanup effort was undertaken that remains a model for the world at large.

The process of smelting involves heating sulfur-containing metal ores in air to convert the sulfide to the oxide form. In the process, the sulfur is removed in the form of sulfur dioxide. Without environmental controls, this gas goes into the atmosphere, where it reacts with water and oxygen to form sulfuric acid. This acid rain damages foliage, lakes, and structures, and causes difficulty for anyone with respiratory ailments. The unnatural acid levels (measured at a pH of less than 3 in some cases) may also lead to the leaching of aluminum from soil into groundwater, which has negative impacts on roots and aquatic life in streams and lakes.

In the late nineteenth and early twentieth centuries, nickel and copper sulfide ores were found in Sudbury, Ontario, Canada, just north of Lake Huron, and a metal mining and processing industry took root there. Extensive logging took place in the region as wood was needed for the growing settlements and for fuel for the smelter operations. Trees were unable to grow back in the deteriorating environment, and the denuded soil quickly eroded into waterways, adding to the devastation. At the same time, emissions from the smelting process included airborne metal particulates that deposited toxic levels of nickel and copper in the surrounding soil and water. By the early 1960's Sudbury was known around the world for its acidified, lifeless lakes and its blackened, treeless landscape.

Late in the 1960's the Canadian government began to respond to the growing worldwide environmental movement by ordering a reduction of sulfur dioxide and metal levels in the air around Sudbury. The ore-processing companies responded by constructing 380-meter (1,247-foot) "superstacks" that acted to reduce local air pollution by spreading it over a larger area. During this period, the world was beginning to recognize the transboundary nature of pollution, particularly air pollution. Dangerously acidic conditions were found in a 50-kilometer (31-mile) radius of Sudbury and even farther in the direction of the prevailing winds.

Citizen outcry made it clear that improvements also had to occur inside the industrial plants. Among the changes implemented were the use of higher-grade ore, recycling of sulfur gases to make and sell sulfuric acid, and treatment of stack gases to remove residual acid and particulates. Sudbury eventually became known for its remediation of local lakes and soil through the application of basic materials (such as lime) and for replanting and nurturing young trees and other plants. Over twenty years, soil and water pH levels rose measurably, and plant and aquatic life slowly began to return to the area. In 1992 Sudbury received commendation at the Earth Summit in Brazil for its unprecedented success in environmental cleanup; the city's efforts were noted as an example from which the rest of the world could learn.

*Wendy Halpin Hallows*

FURTHER READING

Jacobson, Mark Z. *Atmospheric Pollution: History, Science, and Regulation.* New York: Cambridge University Press, 2002.

Keller, W., et al. "Recovery of Acidified Lakes: Lessons from Sudbury, Ontario, Canada." In *Acid Rain: Deposition to Recovery,* edited by Peter Brimblecombe et al. New York: Springer, 2007.

# Sulfur oxides

CATEGORIES: Pollutants and toxins; atmosphere and air pollution

DEFINITION: Chemical compounds containing sulfur and oxygen

SIGNIFICANCE: Sulfur oxides emitted into the earth's atmosphere pollute the air and have negative effects on human and animal health. Sulfur dioxide is the major contributor to acid rain, which causes several types of environmental damage.

Sulfur oxides occur both naturally and as the result of human activities. On a global basis, natural sources, such as volcanoes, contribute about the same amount of sulfur oxides to the atmosphere as do human activities. In industrialized nations such as those of Europe and the Americas, however, human activities contribute 95 percent of the sulfur oxides emitted. The two most important sulfur oxides are sulfur dioxide ($SO_2$) and sulfur trioxide ($SO_3$). $SO_2$ is a colorless, dense, toxic, nonflammable gas with an intense odor; $SO_3$ is a liquid.

In 2005 the U.S. Environmental Protection Agency (EPA) estimated that anthropogenic (human-caused) emissions of $SO_2$ amounted to more than 14 million tons. Of these, 73 percent were the result of electricity generation, 15 percent came from fossil-fuel combustion (coal may contain from 1 percent to 4 percent sulfur, and when burned with oxygen in the air, it produces $SO_2$), almost 8 percent resulted from industrial processes, and smaller amounts came from nonroad equipment, on-road vehicles, fires, waste disposal, residential wood combustion, and solvent use.

$SO_2$ is an air pollutant and a lung irritant. Scientific evidence links short-term exposures to $SO_2$ (5 minutes to 24 hours) with adverse respiratory effects, including bronchoconstriction (tightening of the bronchi in the lungs) and increased asthma symptoms. Studies also show a connection between short-term exposure to $SO_2$ and increased visits to hospital emergency rooms and hospital admissions for respiratory illnesses, particularly in at-risk populations: children (whose lungs are still developing and have elevated breathing rates while playing), asthmatics, and the elderly (who may also have compromised lung function). Sulfur oxides can react with other compounds in the atmosphere to form small particles that can penetrate deep into the lungs and cause or worsen respiratory diseases, such as emphysema and bronchitis, and aggravate existing heart disease.

Acid rain (or acid snow, sleet, or fog) is a direct result of the method that the atmosphere uses to clean itself. Tiny droplets of water in the atmosphere continuously capture suspended particles and gases; $SO_2$ and oxides of nitrogen are then chemically converted to sulfuric and nitric acids. $SO_2$ is the major gas producing acid rain. Both $SO_2$ and $SO_3$ contribute to the formation of acid rain, but $SO_3$ is found in the atmosphere at much lower concentrations than $SO_2$.

Industrial acid rain is a substantial problem in Europe, China, Russia, and areas downwind from them because sulfur-containing coal is burned to generate heat and electricity in these parts of the world. The use of tall smokestacks to reduce local pollution contributes to the formation of acid rain considerable distances downwind of the original emissions by releasing acid-forming gases high into the atmosphere. Acid rain has adverse impacts on forests, freshwater resources, soils, and the human-built environment, killing insects and aquatic life and causing damage to buildings. It also affects human health in ways similar to gaseous $SO_2$.

The EPA's air-quality standard for $SO_2$ is designed to protect against exposure to all sulfur oxides. $SO_2$ is of greatest concern and is used as the indicator for the larger group of sulfur oxides because emissions that lead to high concentrations of $SO_2$ generally lead to the formation of other sulfur

## Global Sulfur Emissions by Source and Latitude

*Atmospheric chemical inputs can vary greatly by region, as illustrated by the table below listing the vastly different sources of atmospheric sulfur in different parts of the globe.*

| LATITUDE | ANTHROPOGENIC % | MARINE % | TERRESTRIAL % | VOLCANIC % | BIOMASS BURNING % |
|---|---|---|---|---|---|
| 90° south | 0 | 0 | 0 | 0 | 0 |
| 75° south | 0 | 80 | 0 | 19 | 1 |
| 58° south | 2 | 97 | 0 | 0 | 1 |
| 45° south | 22 | 72 | 0 | 9 | 1 |
| 28° south | 67 | 28 | 0 | 1 | 4 |
| 15° south | 21 | 47 | 1 | 22 | 10 |
| 0° | 21 | 39 | 1 | 33 | 7 |
| 15° north | 40 | 30 | 1 | 19 | 1 |
| 28° north | 85 | 6 | 0 | 8 | 1 |
| 45° north | 88 | 4 | 0 | 7 | 1 |
| 58° north | 86 | 3 | 0 | 10 | 1 |
| 75° north | 30 | 40 | 0 | 23 | 7 |
| 90° north | 0 | 0 | 0 | 0 | 0 |

*Source:* Pacific Marine Environmental Laboratory, National Oceanic and Atmospheric Administration.

oxides. Annual average ambient $SO_2$ concentrations across the United States have decreased by more than 70 percent since 1980, and by the early twenty-first century all areas of the United States had met the EPA's standard for $SO_2$.

*Bernard Jacobson*

FURTHER READING

Christiani, David C., and Mark A. Woodin. "Urban and Transboundary Air Pollution." In *Life Support: The Environment and Human Health*, edited by Michael McCally. Cambridge, Mass.: MIT Press, 2002.

McKinney, Michael L., Robert M. Schoch, and Logan Yonavjak. "Air Pollution: Local and Regional." In *Environmental Science: Systems and Solutions*. 4th ed. Sudbury, Mass.: Jones and Bartlett, 2007.

Vallero, Daniel. *Fundamentals of Air Pollution*. 4th ed. Boston: Elsevier, 2008.

# Thermal pollution

CATEGORY: Pollutants and toxins

DEFINITION: Adverse environmental effect caused by heat, particularly waste heat from steam-electric power plants

SIGNIFICANCE: When thermal effluent is discharged into waterways from steam-electric power plants, the resulting increases in water temperature can cause damage to the aquatic ecosystems of those waterways.

The overwhelming majority of waste heat in industrialized countries comes not from factories but from steam-electric power plants such as coal-burning and nuclear power plants. Steam-electric power plants convert thermal energy from the combustion of fossil fuels or nuclear reactions into mechanical work and then into electrical energy. While the generators that convert the mechanical work into electrical energy in such a plant are nearly 100 percent efficient, the rest of the plant is subject to maximum efficiencies imposed by the laws of thermodynamics and determined by the highest and lowest temperatures of the plant. Steam-electric power plants typically have efficiencies of about 40 percent or less, which means that 40 percent of the heat is converted into electrical energy, while the other 60 percent becomes unusable waste heat that must be removed.

As an example, consider a large electric power plant producing 1,000 megawatts (MW) of electricity. If its efficiency is 40 percent, the plant will have to produce 2,500 MW of heat (since 1,000 MW is 40 percent of 2,500 MW) to maintain this output. Waste heat will be produced at the rate of 2,500 MW − 1,000 MW = 1,500 MW. At a coal-burning plant, perhaps 200 MW of heat will be lost in and around the boilers and the rest of the plant, leaving 1,300 MW to be removed. Nuclear power plants usually run at lower maximum temperatures, so they have lower efficiencies and correspondingly produce more waste heat; in addition, less heat is lost around the plants themselves, so more of the waste heat needs to be removed. Common methods of disposing of this heat include dumping it into rivers or lakes and or using it to evaporate water in cooling towers. Such disposal methods can have adverse effects on aquatic ecosystems or generate fog and ice.

There are three major methods of removing waste heat. The least expensive is known as once-through cooling, in which water from a stream, lake, or other body of water is used to cool the steam, after which the water is returned to its source at a higher temperature. This method may result in temperature increases of several degrees in the body of water.

A second method is the use of artificial lakes or cooling ponds, which may be up to several square kilometers in size. The heated water from the power plant is discharged into one end of a pond, while water to be used for cooling is drawn from the bottom of the pond at the other end. The water in the pond cools naturally by evaporation; therefore, the pond's water source must be continuously replenished.

A third method is the use of cooling towers, either evaporative or nonevaporative. Evaporative towers, as their name suggests, cool water from the power plant by promoting evaporation. Some evaporative towers produce natural drafts, while others use fans to induce drafts mechanically. Natural draft towers may be more than 100 meters (328 feet) high, while mechanical draft towers are often much smaller. Nonevaporative cooling towers allow moving air to cool pipes containing the heated water from the power plants; these kinds of towers are less commonly used than evaporative towers because they are expensive to build and operate.

Since most cooling methods ultimately lead to the evaporation of substantial amounts of water, large power plants are usually located adjacent to rivers, which provide a source of water. The major ecological effects of thermal pollution occur in natural rivers and lakes and involve fish and other aquatic organisms. These organisms typically thrive when the temperature remains within a narrow range and may die if the water changes to lower or higher temperatures. For example, if a population of largemouth bass that are acclimated to a water temperature of 20 degrees Celsius (68 degrees Fahrenheit) are exposed to temperatures as low as 4.4 degrees Celsius (40 degrees Fahrenheit) or as high as 32 degrees Celsius (90 degrees Fahrenheit) for one or two days, about 50 percent will die. In addition, the sudden changes in temperature that are encountered by fish swimming into thermal effluent can produce thermal shock and almost instantaneous death if the changes are sufficiently large.

All chemical reactions are increased by heat, so thermal pollution can lead to more rapid physiological processes in fish and other aquatic organisms. In certain circumstances this can cause increased growth rates and shorter life spans, leading to decreased populations and less biomass in the ecosystem; in other circumstances it may lead to increased populations and biomass or it may extend the growing system. Ecologists have established that a temperature change of a few degrees can have significant effects, both short-term and long-term, on aquatic ecosystems.

In addition, evaporative cooling methods inject large amounts of water vapor into the atmosphere. During humid weather this may lead to fog, which can cause dangerous visibility issues on nearby roads; during cold weather it may lead to damage by icing roads, trees, and buildings.

*Laurent Hodges*

FURTHER READING

Camp, William G., and Thomas B. Daugherty. "Water Pollution." In *Managing Our Natural Resources.* 4th ed. Albany, N.Y.: Delmar, 2004.

Goudie, Andrew. "The Human Impact on the Waters." In *The Human Impact on the Natural Environment: Past, Present, and Future.* 6th ed. Malden, Mass.: Blackwell, 2005.

Hill, Marquita K. "Water Pollution." In *Understanding Environmental Pollution.* 3d ed. New York: Cambridge University Press, 2010.

Laws, Edward A. "Thermal Pollution and Power Plants." In *Aquatic Pollution: An Introductory Text.* 3d ed. New York: John Wiley & Sons, 2000.

# Thermohaline circulation

CATEGORY: Weather and climate
DEFINITION: Global system of oceanic currents driven by temperature and salinity gradients
SIGNIFICANCE: The global interchange of ocean water moderates climate at high latitudes, oxygenates the ocean depths, and aids fertility and productivity of the marine environment by recycling nutrients.

The existence of surface currents in the ocean has been recognized since antiquity. Some are wind-driven, but others are part of a vast interconnected system whereby warm water from tropical latitudes circulates northward close to the surface, while cold water from the poles sinks to the depths and flows toward the equator.

The driving forces behind this oceanic conveyor belt are differences in salinity and temperature. Surface water that is denser than the water below it, because of either increased salinity or lower temperature, will sink. In temperate-zone lakes, the water simply turns over in winter, but in the ocean, the displaced water flows "downhill" along the latitudinal density gradient. In the Tropics, surface evaporation increases the salinity, and therefore the density, of surface waters. In certain areas, particularly the northeastern Pacific Ocean and along the west coast of Africa, this leads to upwelling of nutrient-laden waters that support high oceanic productivity. The effect is too diffuse to enhance or retard the conveyor itself. Elevated temperatures may produce a "sluggish" conveyor, but the overall pattern has remained stable for at least the past fifty million years.

The cold, dense water driving thermohaline circulation forms principally in two regions: in the North Atlantic off the coast of Greenland and in a band encircling Antarctica. In the North Pacific, the narrow Bering Strait and shallow waters surrounding it prevent the southward flow of cold, dense water. In the Pacific, a deep current originating in Antarctica flows northward in the western Pacific to middle latitudes, while a southerly surface current loops around the west coast of North America, crosses the Pacific near the equator, passes north of Australia, merges with northerly surface currents of Antarctic origin as it rounds Africa, and continues northward along the western Atlantic margin and the east coast of Greenland.

## Thermohaline Circulation

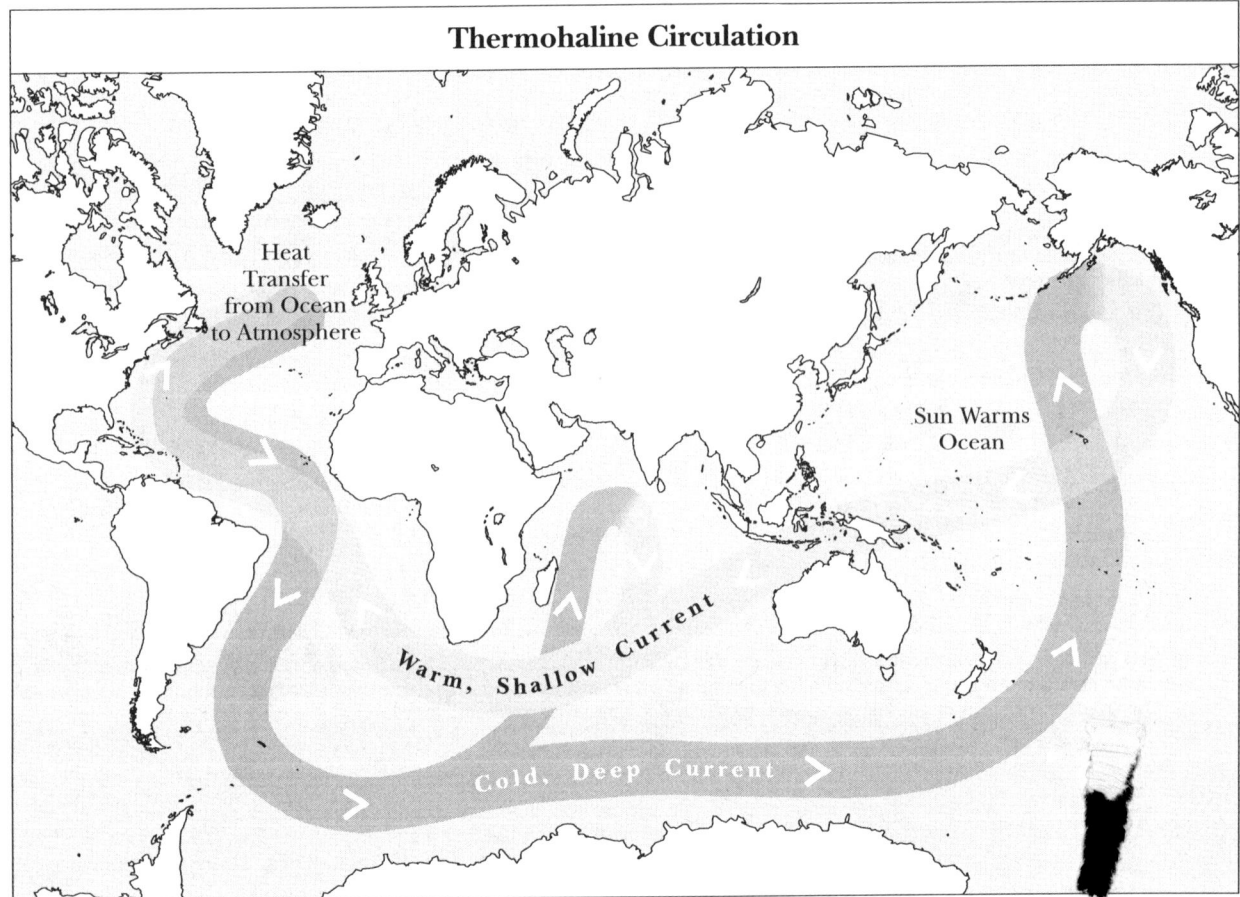

Salinity and the temperature of surface currents both increase in the Tropics. In the present climate regime, these effects cancel each other out. To the extent that changes in global temperature affect the action of the conveyer, that effect is much more profound in polar regions.

Cold temperatures increase both the density and the salinity of seawater. Because sea ice is effectively fresh water, salinity beneath it increases. Any large-scale melting of polar sea ice can therefore be expected to reduce the rate of formation of cold water and thus slow the circulation of water in the oceanic conveyor belt. How much this would affect climate is uncertain. A widely held theory concerning a cold period, the Younger Dryas, and the end of the last ice age links it to rapid influx of fresh water into the North Atlantic and disruption of warm surface currents. It is uncertain whether a more gradual process would have a comparable effect.

In addition to affecting climate, the thermohaline circulation carries oxygenated water into the ocean depths, supporting life in the deep sea and preventing the buildup of organic matter under anoxic (oxygen-deprived) conditions on the ocean floor. Any significant disruption of these currents would, therefore, have undesirable implications for the diversity and productivity of marine life.

*Martha A. Sherwood*

FURTHER READING

Aken, Hendrik Mattheus van. *The Oceanic Thermohaline Circulation: An Introduction.* New York: Springer, 2007.

Marshall, John. *Atmosphere, Ocean, and Climate Dynamics: An Introductory Text.* Boston: Elsevier Academic Press, 2008.

Oliver, John E., ed. *Encyclopedia of World Climatology.* New York: Springer, 2005.

# Volatile organic compounds

CATEGORIES: Pollutants and toxins; atmosphere and air pollution

DEFINITION: Broad class of natural and human-made organic compounds that evaporate readily and have low solubility in water

SIGNIFICANCE: In indoor environments high concentrations of volatile organic compounds emitted by household products and construction materials can lead to reduced air quality and potential health problems. In urban and industrial environments these compounds play an important role in the development of tropospheric ozone and photochemical smog, and they have also been shown to play a role in groundwater contamination, the depletion of stratospheric ozone, and global climate change.

Volatile organic compounds (VOCs) are highly reactive pollutants that are emitted by both human-made and natural sources. The definitions as to what constitutes a VOC vary among scientific organizations and regulatory agencies in different countries. In general, however, VOCs are characterized as organic, or carbon-based, compounds that have a high vapor pressure, meaning that they evaporate easily under normal atmospheric conditions. Abundant organic compounds such as carbon monoxide ($CO$), carbon dioxide ($CO_2$), and methane ($CH_4$), along with other less reactive compounds, are normally not considered to be VOCs.

VOCs are typically present in the atmosphere in trace quantities. Plants and trees are the largest natural sources of VOCs; however, they are also emitted by soil microbes, forest fires, animals, and the oceans. Anthropogenic sources—that is, those related to human activities—include the use of liquid and gaseous fossil fuels for transportation and power production, biofuel burning for heating and cooking, and agricultural pesticide use. VOCs are also emitted by ingredients that are commonly used in building materials and such household products as cleaning supplies, paints, cosmetics, and disinfectants.

Natural sources account for much more of total VOC emissions than do anthropogenic sources. The impacts of the natural emissions on air quality, however, are minimized by the fact that natural sources are widely distributed across space. Human-caused emissions, in contrast, often have far greater impacts on air quality because they are typically concentrated in urban and industrial regions or, at a smaller scale, inside buildings. In urban regions VOCs, particularly unburned gasoline or hydrocarbons, react with nitrogen oxide ($NO$) and nitrogen dioxide ($NO_2$) in the presence of sunlight to form tropospheric ozone ($O_3$); VOCs and ozone are primary ingredients in the formation of photochemical smog. By modifying the concentration of ozone ($O_3$), a greenhouse gas, VOCs not only pose a health concern in urban regions but also have an indirect effect on global warming. In addition, the improper disposal of VOCs, including gasoline and industrial solvents, can contaminate groundwater and adversely affect drinking-water sources.

In indoor environments VOCs pose a particular health concern because their concentrations may build to higher levels than are typically found outdoors. The widespread presence of VOCs in household products used indoors has the potential to expose people to a variety of VOC compounds at relatively high concentrations for long exposure periods.

Scientific and regulatory efforts to identify and describe the health impacts of VOCs are complicated by the fact that many different types of VOCs are released into indoor and outdoor environments and their impacts are quite variable; some VOCs have no health effects, whereas others are known cancer-causing agents. Because of this variability, efforts to regulate VOC emissions are diverse. Notable efforts to reduce VOCs have included the development of the catalytic converter, the regulation of toxic emissions under the U.S. Clean Air Act, and the regulation of municipal and industrial waste handling, drinking-water quality, and indoor air quality by the Environmental Protection Agency.

*Jeffrey C. Brunskill*

FURTHER READING

Ahrens, C. Donald. *Meteorology Today: An Introduction to Weather, Climate, and the Environment.* Belmont, Calif.: Brooks/Cole Cengage Learning, 2009.

Girard, James E. *Principles of Environmental Chemistry.* Sudbury, Mass.: Jones and Bartlett, 2005

Koppmann, Ralf, ed. *Volatile Organic Compounds in the Atmosphere.* Hoboken, N.J.: Wiley-Blackwell, 2007.

# Windscale radiation release

CATEGORIES: Disasters; nuclear power and radiation

THE EVENT: Release of radioactive material into the atmosphere as the result of a fire in the reactor core of the Windscale nuclear reactor on the west coast of England

DATE: October 10, 1957

SIGNIFICANCE: The fire in the reactor core at the Windscale plant gave rise to one of the world's first serious nuclear accidents. The release of significant amounts of radioactive material into the atmosphere caused short-term contamination of several hundred square miles of the surrounding countryside.

The Windscale nuclear reactor overheated in October, 1957, and a fire resulted in the reactor core. The first indication that radioactive material was escaping into the atmosphere from the reactor came on the evening of October 10, when a nearby weather station detected an increase in background radiation.

Health physicists considered the release of iodine 131 to be the most serious hazard. Iodine 131 falls to the ground, where it may be consumed by cows eating grass and concentrated in their milk. If humans drink this contaminated milk, the iodine 131 concentrates in the human thyroid gland, where its radioactive decay can cause cancer. The British government monitored milk from the region for evidence of iodine-131 contamination, and two days after the first release of radioactive material, milk samples from farms near the Windscale plant showed evidence of contamination. Initially, the government impounded milk supplies from within a 3-kilometer (2-mile) radius around the plant. However, as iodine-131 contamination was detected over a wider region, milk produced over an area of about 500 square kilometers (200 square miles) was impounded. The contaminated milk was dumped into the sea, and milk for the people living near the Windscale reactor was trucked in from outside the contaminated region. Since iodine 131 decays rapidly, the ban on consumption of milk from the affected area lasted only a few weeks.

The fire in the reactor core also released a significant amount of polonium 210 into the atmosphere.

This raised concerns because polonium 210 decays by emitting alpha particles, which are dangerous to the lungs. However, in the areas where the concentration of polonium 210 was highest, the additional exposure to radioactive decay was found to be approximately equivalent to the average annual background rate of radioactive decay in the British Isles.

The design of the Windscale nuclear plant minimized the public health hazard posed by the reactor fire, as filters in the stacks of the plant trapped a large fraction of the radioactive material released from the reactor. The quick action of the British government in collecting and destroying contaminated milk from the affected region also reduced the health effects of the release.

The Windscale event released about 0.001 times the amount of iodine 131 into the atmosphere that was released by the fire at the Chernobyl nuclear reactor in 1986. A study conducted in 1997, forty years after the Windscale release, concluded that individuals who received the most serious exposure to the Windscale radiation experienced a slight increase in their likelihood of developing fatal cancers compared to the normal fatal cancer risk. The conclusion was that the long-term health effects of the Windscale release were minimal. In 2007, however, British scientists announced the results of a new study that used computer modeling and environmental monitoring to examine how the radioactive materials released would have spread. They concluded that the environmental contamination caused by the Windscale release was probably greater than originally thought. Whereas previous estimates had put the number of cases of cancer eventually caused by the radiation at 200, the results of the study suggested that a more accurate estimate would be 240 cases.

*George J. Flynn*

FURTHER READING

Bodansky, David. "Nuclear Reactor Accidents." In *Nuclear Energy: Principles, Practices, and Prospects.* 2d ed. New York: Springer, 2004.

Cooper, John R., Keith Randle, and Ranjeet S. Sokhi. "Nuclear Power." In *Radioactive Releases in the Environment: Impact and Assessment.* Hoboken, N.J.: John Wiley & Sons, 2003.

# Yokkaichi, Japan, emissions

CATEGORIES: Disasters; atmosphere and air
  pollution
THE EVENT: Heavy pollution of the air in the Yokkaichi area, caused by petrochemical plants
DATES: 1950's-1970's
SIGNIFICANCE: The industrial pollution of the air in
  and around Yokkaichi, Japan, caused widespread
  health problems for residents and led to landmark
  court cases and legislation aimed at preventing and
  reducing air pollution.

Yokkaichi, a port city on Japan's Ise Bay, developed
as a major industrial and petrochemical center in
the early twentieth century. The demands of World
War II and Japan's postwar recovery led to further industrial expansion in the area, and an oil refinery
complex known as the Yokkaichi Kombinato was created in the 1950's. Although the complex was an economic success, the pollution it generated was soon
linked to breathing difficulties and a variety of other
health problems in area residents. Researchers found
a high correlation between airborne sulfur dioxide
and the incidence of bronchial asthma in children
and of bronchitis in older people. Nevertheless, in
1963, a second industrial complex was opened in the
region, and a third was added in 1973. In one district
of Yokkaichi, airborne sulfur dioxide levels were
found to be 800 percent above normal. In the early
1960's, nearly one-half of the area's young children,
nearly one-third of its elderly, and approximately one-
fifth of its young adults had developed respiratory ab-
normalities.

In 1967 a group of Yokkaichi residents filed a suit
against the Shiohama Kombinato, which ran one of
the petrochemical complexes, and in 1972 the plain-
tiffs were awarded nearly $300,000 in damages. The
award marked the first time that a group of Japanese
companies had been held liable for damages, setting
a precedent that made other companies vulnerable to
such litigation. As a result of the case and ensuing controversy, in 1967 Japan's government enacted a basic
antipollution law. Within the next several years, additional laws spelled out redress rights for victims from
the Yokkaichi area and for residents of polluted areas
near Kawasaki and Ōsaka. Regulations requiring refineries to adhere to pollution-abatement policies
were also strengthened.

As a result of such measures, by the mid-1970's airborne sulfur dioxide levels in the Yokkaichi region
had decreased more than 60 percent, and the rate of
respiratory complaints among area residents had declined sharply. By the 1990's nearly 100,000 Japanese
citizens had been declared eligible for compensation
under the new laws.

*Alexander Scott*

# Bibliography

Burroughs, H. E., and Shirley J. Hansen. *Managing Indoor Air Quality.* 4th ed. Lilburn, Ga.: Fairmont Press, 2008.

Finlayson-Pitts, Barbara J., and James N. Pitts. *Chemistry of the Upper and Lower Atmosphere: Theory, Experiments, and Applications.* San Diego, Calif.: Academic Press, 2000.

Griffin, Roger D. *Principles of Air Quality Management.* 2d ed. Boca Raton, Fla.: CRC Press, 2007.

Jacobson, Mark Z. *Atmospheric Pollution: History, Science, and Regulation.* New York: Cambridge University Press, 2002.

Kessel, Anthony. *Air, the Environment, and Public Health.* New York: Cambridge University Press, 2006.

Lane, Carter N., ed. *Acid Rain: Overview and Abstracts.* New York: Nova Science, 2003.

Lipton, James P., ed. *Clean Air Act: Interpretation and Analysis.* New York: Nova Science, 2006.

McCarthy, Tom. *Auto Mania: Cars, Consumers, and the Environment.* New Haven, Conn.: Yale University Press, 2007.

Seinfeld, John H., and Spyros N. Pandis. *Atmospheric Chemistry and Physics: From Air Pollution to Climate Change.* 2d ed. Hoboken, N.J.: John Wiley & Sons, 2006.

Simpson, John A., ed. *Preservation of Near-Earth Space for Future Generations.* New York: Cambridge University Press, 2007.

Sokhi, Ranjeet S., ed. *World Atlas of Atmospheric Pollution.* London: Anthem Press, 2008.

U.S. Environmental Protection Agency. *The Plain English Guide to the Clean Air Act.* Research Triangle Park, N.C.: Author, 2007.

# CATEGORY INDEX

# INDEX